R.A.Whitehead

GARRETT STEAM TRACTORS & ROLLERS

R.A.Whitehead & Partners
Tonbridge,
Kent

1999

On a glorious pre-war June day Bob Pratt caught the scene everyone, except possibly the engine crew, would wish to relive as tractor No.33739, built in 1920 for the Eddison Steam Rolling Co. Ltd. of Dorchester but by then owned by the famous West of England showmen J. Rowland & Son of St. Austell, hauled its loads through a village in Devon. [R.G. Pratt]

Published by R.A.Whitehead & Partners
42 Hadlow Road
Tonbridge, Kent TN9 1NZ

Photographs: Unless otherwise acknowledged all illustrations are from the collections of the now defunct firm of Richard Garrett Engineering Ltd., or of the author. Illustrations from other sources are acknowledged individually.

Cover: Noah Etheridge, in one of his famous Stetson hats, poses with his newly acquired tractor, No.27886, under the wintry elms surrounding the stack-yard at Blythburgh Hall in the Winter of 1909/10, the whole plant - tractor, thresher, and straw elevator - made by Garretts.

Typeset in Times New Roman 10 by J.E.Whitehead.

Printed and bound by Biddles Ltd., Woodbridge Park, Guildford, GU1 1DA

ISBN 0-9508298-9-7

Contents

Foreword by J.W. Powell, Esq.		Page 5
Preface		Page 6
Acknowledgements and thanks		Page 6
Chapter 1	Heavy Rollers 1898-1909	Page 11
Chapter 2	Why Build Tractors?	Page 19
Chapter 3	The Three Ton Tractors	Page 25
Chapter 4	The Five Ton Tractors 1906-1911	Page 36
Chapter 5	The Three-Shaft Rollers	Page 50
Chapter 6	The Later Heavy Rollers 1910-1919	Page 57
Chapter 7	The Later Tractors 1912-1919	Page 67
Chapter 8	The DR Tractors	Page 75
Chapter 9	Under the AGE regime	Page 89
Chapter 10	Uses and Users	Page 104
Chapter 11	Characteristics & Preferences	Page 113
Specifications of :		
	Standard Steam Road Rollers	Page 125
	Standard Steam Motor Tractors	Page 129
Appendix 1	Table of Dimensions of Tractors & Rollers c.1909	Page 132
Appendix 2	Telegraphic Codes	Page 133
Appendix 3	Key to Prefixes Applied to Drawings & Patterns	Page 133
Appendix 4	Indexes of Garrett Steam Tractors and Rollers	Page 134

Fig.1. *One of the first ten rollers made, probably No.23501, finished and taken into the stockyard by November, 1901, and sold to William Rawlings of Collingbourne Ducis at the end of March, 1902. This photogaph shows the engine in primer during its period in stock. In these first rollers the boiler barrel remained the same diameter as in the 6NHP traction engine but was reduced in length to 4' 6½" in lieu of the original 5' 8". I once inadvertently coupled this picture to a caption stating it to be No.34253, a misdemeanour which I doubt if I shall ever live down. The roller has the early type of dead-weight scrapers. The distinctive easing arm of the Ramsbottom type safety valve was a feature in common with the contemporary Garrett traction engines.*

Fig.2. *Garrett 4CD No.32048 (1914)* Lady Raglan *en route to Ashton Keynes early in 1953, after purchase by Joseph Powell.* [J.W. Powell]

Foreword

by J.W. Powell, Esq., Past President of the Road Locomotive Society

When Bob Whitehead asked me to write a Foreword for his latest book *Garrett Steam Tractors & Rollers* I was delighted to have been asked by someone who has written so many interesting books on the products turned out at Leiston Works since 1778. I first saw the name of Richard Garrett & Sons, when still very young, in the 1851 edition of the Illustrated London News devoted to the Great Exhibition; this volume was bound by my maternal great grandparents and handed down to my mother who passed it on to me. I well remember the many engravings of steam engines, agricultural machinery, and Garretts' impressive stand with artistic columns capped with urns, plus a coat of arms.

My father's family were farmers and used contractors to do the steam ploughing, threshing, etc. In 1917 he foresaw that the future lay in paraffin tractors, so went into agricultural engineering, although doing steam repairs when asked. As a matter of interest the rate for threshing was £2 a day, ploughing £1 an acre, and cultivating 10/- per acre. It took time for the farm workers to adjust to driving tractors and to learn not to overstretch them; they needed many repairs - dust in the bearings and, commonly, the connecting rod coming out through the crankcase because of overspeeding.

My first job at the age of 14 was with steam, helping to retube a 12HP Robey portable loco boiler. In 1951 I bought my first Garrett tractor, No.33738, built 1919. This came from W.A. Rawlings of Collingbourne Ducis, and had been worked hard in the past on timber hauling. It has been completely rebuilt with the help of drawings and spares supplied by Garretts. In 1953 I acquired another Garrett tractor, No.32048 (1914), which I ran for several years, but am now rebuilding. My task has been greatly helped by these engines being so well designed, built to stand up to hard work and last for a long time. This is proved by the number that are now in preservation, and we are lucky to have so many enthusiasts to ensure their survival.

Thus Garretts have always played an important part in my life, and we should all be grateful to Bob Whitehead for his research and recording over many years, and for managing to put together this book of such interesting and important material gathered from records before the final closure of Leiston Works.

Joseph W. Powell
Ashton Keynes. January, 1999

Preface

Because only thirty-nine Garrett steam rollers were sold new in the British Isles, many of them in Ireland, Garretts are frequently dismissed as being insignificant in that field whereas the contrary is true. Although Aveling & Porter rollers outsold all other makers, Garretts, with nearly four hundred rollers to their credit, were entitled to rank with Fowlers, Marshalls, and Wallis & Steevens, and this despite the deliberate snuffing out of their roller production in favour of Rochester products under the AGE regime.

On the other hand their steam tractors, or more particularly the No.4 compound tractors, have been accorded a generous measure of appreciation. In the brief period from 1907 until 1929 five hundred and fifteen 4CDs were made even though after 1919, under the AGE control, they were built only under sufferance, so to speak, since group policy had allocated the production of steam tractors to Aveling & Porter. As it turned out customer preferences kept the Garrett tractor going for some years more although in the last four years only fifteen were made.

Notwithstanding this shading-off in the final years more tractors were sold than traction engines and along with the overtype steam wagons, whose story was told in *Garrett Wagons - Part 1*, in the run-up to the 1914-18 war they gave their builders the much desired entrée into the home market.

The Garrett trade in steam engines was dominated by the portable. Out of a total of some 22,500 engines made, $2\frac{3}{8}\%$ were traction engines or road locomotives, $2\frac{1}{2}\%$ tractors, $1\frac{5}{8}\%$ rollers, and $4\frac{1}{2}\%$ wagons. The remaining 89% represented portables, semi-portables, and fixed engines, the bulk of which - probably more than four-fifths - were for export. Of self-movers in all categories (tractions, rollers, tractors, and wagons) 32% were sold overseas, but this was mainly the result of the high percentages of traction engines (49%) and rollers (90%) that went abroad. The collapse of the trade in portables and the loss of export markets after World War 1 were a combined catastrophe that brought Leiston Works to its eventual closure.

Tonbridge, Kent
December, 1998

Robert A.Whitehead

Acknowledgements and Thanks

The assembling of the information for this book was begun well over thirty-five years ago. Because of this it is inevitable that a large number of those whom I have to thank must be referred to in the past tense. The most senior of these was the late Victor R.Garrett, the last family director of the Works to have survived, and an invaluable source of information on the reasons behind directorial decisions. I am also greatly indebted to those who successively had charge of the Works during the period of my researches: the late Alfred R.King, Reginald J.Hadfield, and Michael Hilton. Other directors who gave valuable help were Frank Andrews, Ernest Cuthbert, and Reginald Clarke, all now dead, the latter of whom became a valued personal friend.

The drawing office staff were also very helpful, especially Leslie W.Farrow, chief designer for much of this time, and his deputy, Arthur Woodhead. The late Herbert R. (Jack) Simpson gave much help as did the late Arthur Pipe. The late Edward C.Dunn, a friend for many years, gave immense help both in locating records and documents, and in personal reminiscences of some of the people involved. George Hambling, Walter Hall, and Geoffrey Marreco, all now dead, contributed a great deal of practical information.

Others who have been very helpful are the staffs of the National Motor Museum, Beaulieu; the Rural History Centre, Reading; the Library of the Institution of Mechanical Engineers; the County Records Office of Suffolk County Council; the Long Shop Museum, Leiston, particularly David Williamson who has since moved on to take over the management of the *Buckie Drifter* on the Moray Firth.

Valuable help was given by my son-in-law, Michael J.Walters, who did much research, and also read, commented upon and contributed to the manuscript. He prepared the list of vehicles in the Appendices. My wife, Jean, contributed a share beyond measure and without complaint to the twenty years or so of research we put in at Leiston works and about fifteen years elsewhere, besides editing the manuscript and setting it into type. Over the years Barry Finch gave much help in copying old photographs. Throughout this series Alan Martin has allowed me to use photographs from his vast collection. Pat Freeman has done a magnificent job of proof reading.

To all those named above I extend my heartfelt and most sincere thanks as well as to all those who have contributed but whom consideration of space precludes from individual mention.

In conclusion I must say how grateful I am to Joe Powell for contributing the foreword. He was a pioneer preserver of a Garrett No.4 tractor and, moreover, is able to recall the days when examples were in day-to-day commercial use in the hands of such owners as the firm founded by the late William ('Boiler') Rawlings.

Bibliography

The information contained in this book was, in the main, extracted from documents that existed at Leiston Works. Other sources that were consulted included the periodicals:

Commercial Motor | *Engineer* | *Implement & Machinery Review*
Motor Traction | *Engineering*

Other books by R.A.Whitehead

¶ denotes still in print

The Story of the Colne Valley (jointly with F.D.Simpson)	Oakwood Press [Reprint]
Garretts of Leiston	Percival Marshall
A Century of Service	Eddison Plant
The Age of the Traction Engine	Ian Allan
A Century of Steam Rolling	Ian Allan
Steam in the Village	David & Charles
Garrett 200	Transport Bookman
Kaleidoscope of Steam Wagons	Marshall Harris & Baldwin
Kaleidoscope of Traction Engines	Marshall Harris & Baldwin
Wallis & Steevens - a History	R.L.S.
¶ *A Review of Steam Tractors*	R.L.S.
¶ *Austrian Steam Locomotives 1837-1981*	R.A.Whitehead & Partners
¶ *The Beloved Coast & the Suffolk Sandlings*	Terence Dalton
¶ *Jesse Ellis & the Maidstone Wagons*	R.A.Whitehead & Partners
Steam is the Essence	R.A.Whitehead & Partners
¶ *Garrett Diesel Tractors*	R.A.Whitehead & Partners
¶ *Made by Garretts*	R.A.Whitehead & Partners
¶ *Garrett Wagons Part 1 - Pioneers & Overtypes*	R.A.Whitehead & Partners
¶ *Garrett Wagons Part 2 - Undertypes*	R.A.Whitehead & Partners
¶ *Garrett Wagons Part 3 - Electrics & Motors*	R.A.Whitehead & Partners
¶ *Garrett Traction & Ploughing Engines*	R.A.Whitehead & Partners

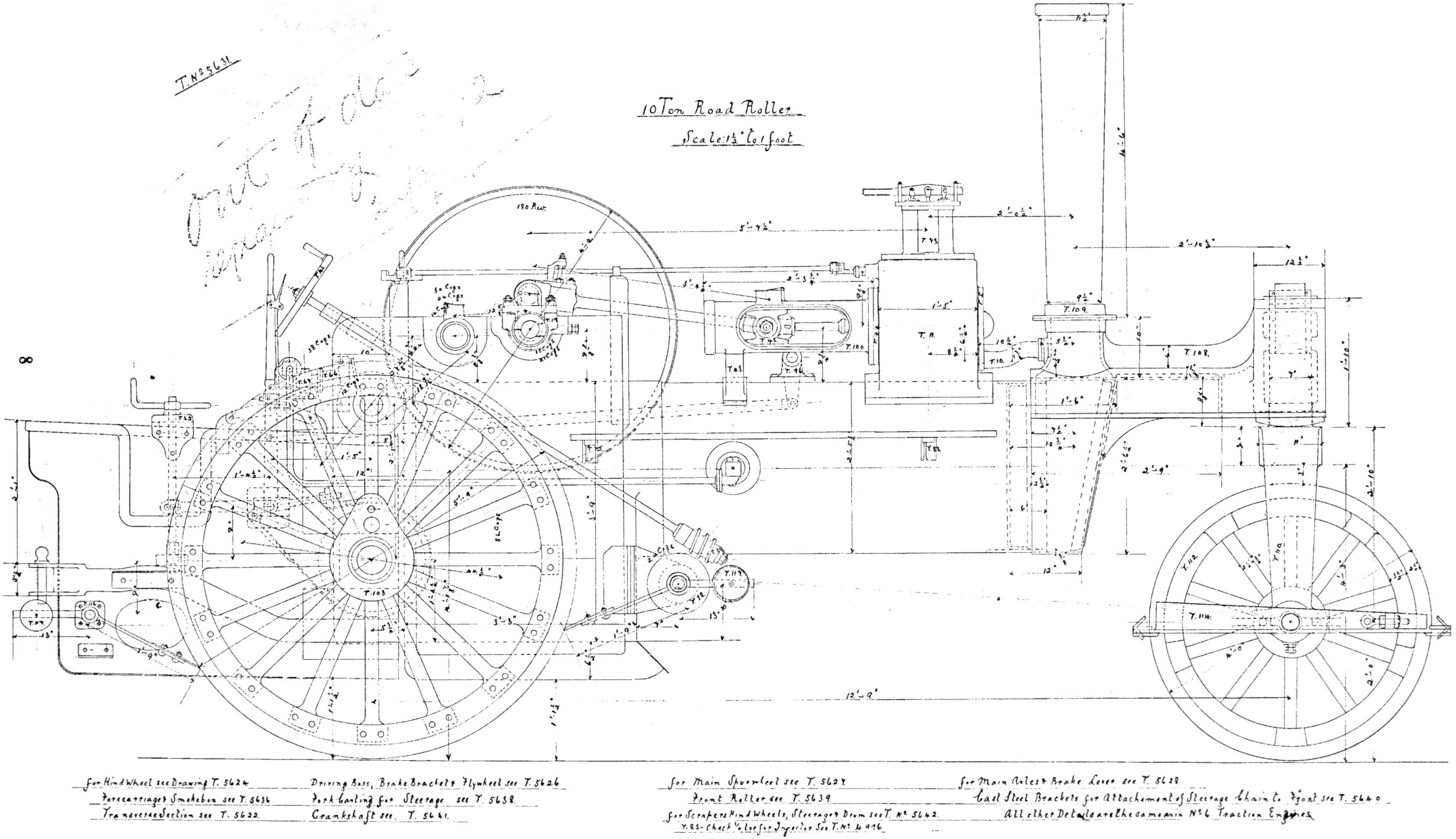

***Fig.3.** The drawing from which the first road roller, No.21411, was constructed using the same boiler and engine work as in the 6NHP traction engine - regrettably the original is in a somewhat damaged condition.*

Fig.4. A standard 10-tonner, supplied to Burn & Co. of Howrah, at work in India. Neither the location nor the number of the roller have been identified. The roof is locally made.
[A.J. Martin collection]

Fig.5. 15 ton compound No.24468, (August, 1903) No.4 in the fleet of Belfast Corporation, unusual, for a home order, in having spring balance safety valves. In April, 1946, an enquiry was received for a new firebox for it, by then owned by A.H. Jordan, Belfast.

Fig.6. Another early 10-tonner, No.24526 (October, 1903), new to W.R. Johnson of Wendling, Norfolk, but when photographed owned by Percy Parsons of Epsom. The picture was taken at Skeet Hill, Orpington, on May 12, 1946. Note the parallel replacement chimney.
[The late B.D. Stoyel]

Fig.7. *10 ton single No.23964 (October, 1902) in the hands of A. Bailey of Wymondham, Norfolk. In 1906 it came back to Garretts, by whom it was reconditioned and sold to Henninger in Germany.*

Fig.8. *Works view of a 10 ton compound roller, not identified with absolute certainty but in all probability No.25424 of August, 1905.*

Fig.9. *The 8 ton single cylinder roller No.27145, sent in May, 1908, to H. Boesewinkel of Zutphen, Holland, ordered with wider rear rolls for work on embankments. This was a slightly lighter version of the* Reading *design.*

Chapter 1

Heavy Rollers 1898 - 1909

In *Garrett Traction & Ploughing Engines* I have related how the arrival of the two elder sons of Frank Garrett, Senior, on the Leiston Works scene began the effort to wean the company away from its all but total reliance upon portable engines and the threshing machine as staple products. Despite their father's aversion to change they persuaded him into a resumption of traction engine building in 1894/95, albeit with distinctly limited sales successes. The engine then designed and made was a 6NHP single cylinder agricultural traction. After further persuasion of their reluctant parent, Frank, Junior, and Alfred Garrett succeeded in securing his agreement to making this the basis of a design of 10 ton single cylinder roller. The older man's reluctance to embark upon the project is understandable. Up to that time the Works had never designed or built a road roller, a field of production in which Aveling & Porter's undeniable experience gave them a tremendous advantage in selling. Probably only the fact that the experiment could be launched with the minimum of capital outlay swayed him into giving his consent.

For something like two decades the firm had been trading with Burn & Co. of Howrah, near Calcutta, who were importers of machinery and engineers' requisites to the Indian sub-continent. Burns, for instance, had purchased Garrett winding engines and winches from Leiston Works in the 1880s. An enquiry from Burn & Co. for a steam roller came at an opportune moment though it is not now clear whether its coming enabled the two younger Garretts to win over their father or whether they had already secured his consent and had cast around for potential customers, gaining this order as a result. Up to the time that the Burn & Co. order was received the sons' position had been somewhat unenviable in that the 6NHP traction engine, the building of which they had canvassed so vigorously, had failed signally to make any impression upon the market.

By using the engine work of the 6NHP traction engine with its 8" diameter x 12" stroke cylinder and its four-shaft arrangement new design was kept to a minimum. Indeed the limited number of new drawings and patterns required for the roller used the same prefix 'T' as was used for the traction engine. In the forks and forecarriage head there was strong evidence that the draughtsman had taken a lead from what was done on the Aveling & Porter 10-tonner. The boiler of the traction engine was retained though the crankshaft, gearing, and rear axle were revised. The smokebox was smaller than on the traction, being 12" wide at the bottom and 1' 6" at the top. The gear side elevation and cross section of this roller were shown on drawing T5631, dated February 3, 1898. The rear wheels were 5' 9" diameter x 1' 4" wide, and the fronts 4' 0" diameter x 2' 1½" wide. Like most of its contemporaries the roller weighed in at over 12½ tons empty and about 13¼ tons in working order.

At Burn's request the crankshaft was made long enough to take a pulley, though the pulley itself was not fitted, and the disc flywheel was 4' 2" in diameter x 6" on the face, suggesting that belt work was intended. On the other hand it was specially mentioned on the order that governors were not to be fitted. Later rollers, however, had a 5" flywheel. No.21411, the first roller to be put in hand, was erected by Walter Rogers, a chargehand often entrusted with new or experimental work, beginning in May, 1898. It left the Works on June 2, 1898, to be followed on August 24, by No.21494, also destined for Burn & Co. at Howrah. The draughtsman principally responsible for the additional drawing work was Arthur Girling. Arthur, the son of Ishmael Girling, foreman of the Thresher Department, had come into the drawing office as a lad and therefore had acquired his engineering knowledge from Emerich Schmach, the chief draughtsman. This background provided him with sound instruction and gave him considerable insight in interpreting Schmach's ideas and wishes but denied him direct experience of how such matters were transacted in other firms. The work on the 10 ton roller in the Winter of 1897/98 probably represented his first major project. Notwithstanding that the gear side elevation and the cross section, T5631, dated only from February 3, 1898, they were soon superseded by T5872 and T5809. The latter was dated March 26, 1899, but the former has lost its date. On January 3, 1900, a third roller, No.22393, again identical with 21411, was sent to Burn & Co. to be followed by a fourth of the same design, No.23039, sent out on December 3, 1900. Whilst we do not have the names of the ultimate customers for the first three rollers it is known that No.23039 was supplied to the Municipality of Mandalay where it was still at work in 1951. In that year a letter of enquiry was received at Leiston, written by the chief officer of the Mandalay Fire Brigade, asking for a quotation for the supply of a similar roller to replace No.23039 which, though by then worn out, had been so satisfactory. Since all stocks of parts for these rollers had long been exhausted Richard Garrett Engineering Works Ltd., the then owners of Leiston Works, felt obliged to decline the invitation to quote. As can be seen from the index of rollers in Appendix 1, Burn & Co. continued to sell rollers of this type at intervals, the next being No.23476 for Cawnpore Municipality sent on August 1, 1901. An Allen 3-tyne scarifier for this was sent out in March, 1916. After these four rollers had been put in hand for Burn & Co. it seems that the design was revised a little. Dating the revisions is not wholly possible but T5872 which superseded the original gear side elevation (T5631) was dated March 22, 1899, so that its predecessor had a currency of less than a year. In it the rear rolls were altered to 5' 8½" diameter x 1' 7" wide, giving better surface pressures. The boiler was shortened by 13½" to 4' 6½", the dimension from the centre of the cylinder to the centre of the crankshaft coming down from 5' 7¼" to 4' 8¾" necessitating a connecting rod only 2' 6" between the pin centres. Subsequently this was lengthened to 2' 7½" and the cylinder was moved closer to the chimney by 1½". One assumes that the original wheelbase of 12' 9" was found to be too long and the changes in the revised design decreased it to 11' 6½", aided in part by reducing the smokebox mean width from 15" to 11½".

With his customary vigour Carl Remy, the German manager of Continental sales, had begun canvassing orders for the new rollers. His first success was with W. & J. Scheid of Limburg a.d. Lahn in Germany, who ordered a pair for delivery without the rear rolls which were to be made in Germany thus, they anticipated, enabling them to avoid the heavy protective tariff applied to finished machinery imports. These two rollers, Nos.23147 and 23376, were despatched on April 4 and May 4, 1901, respectively. In April, 1906, No.23147 needed a new crankshaft, but otherwise nothing is known of their later history. Both were pressed to 11 atmospheres and each had a screw-down brake fitted on the rear axle. It was not long before the Scheids ordered a further 10-tonner of the same design which, numbered 23546, left the Works on March 14, 1902.

In 1901, since roller sales in the home market had still evaded the firm, it was decided that No.23377 (June 10, 1901) should be given show finish and exhibited that year at the Royal Counties Show. This strategy produced results as the workmanlike appearance and good finish of the roller so impressed William Rawlings, of Collingbourne Ducis, Wiltshire, that he bought it. The name that he gave it, *General French,* was that of one of the Boer War generals. At or soon after the show he ordered a second roller which, as No.23501, left Leiston on March 29, 1902, named after *Lord Kitchener* who had become a household name during the relief of Khartoum and had achieved further prominence in the Boer war. Rawlings had a third roller of the same design, No.23745, on June 30, 1902, and this he named *Highwayman.* Wiltshire County Council made it their practice to hire, rather than to own, the steam rollers used on the county's roads and this, in conjunction with the move towards road improvement that began with the new century, had provided a niche for the hiring out of rollers that William Rawlings had noted and was able to help to fill. All three rollers had long lives, the first and third working until about 1950.

Meanwhile further progress had been made in sales to Germany. A. Henninger & Co. of Darmstadt, who had an extensive trade in Garrett portables and threshing machines, and had bought the first 6NHP traction engine made when manufacture was resumed in 1895, enjoyed a very good relationship with Remy and the principals of the Leiston firm. Somewhat later, c.1905-1908, his son was at Leiston Works as what was known as a 'gentleman apprentice'. Although roller No.23486, supplied to Henninger Senior on September 19, 1901, was ordered as a 10/12 tonner, in fact it hardly differed from the standard design other than having boiler fittings to suit German boiler law. Unlike the rollers for the Scheids it seems to have been supplied complete though later rollers sold to Henninger conformed to the practice of other German purchasers in having rims and sundry other parts added in Germany.

Two points emerged from the trade with Germany. The first was that when pressed to 11 atmospheres, as was the case with the trio supplied to Scheids, the rollers were overprovided with steaming capacity and, secondly, notwithstanding that the standard 10 ton (nominal) roller weighed in excess of 13 tons when in working order, there was a demand, albeit of limited extent, for still heavier rollers. The 10-tonner was amended once more with a revised design of boiler, in which the length of barrel was further reduced to 4' 3³⁄₁₆" with the diameter cut from 2' 5½" to 2' 1½". This in turn, of course, altered the saddle base of the cylinder. At the same time the tube size was reduced from 2" to 1¾", though two additional tubes were used viz. 34 as against 32 as formerly. The gear side elevation showing these changes, No.6312, prefixed 'ST' instead of 'T' as previously, was dated February, 1901. Increasing the nominal weight from

10 tons to 12 tons, when so ordered, was accomplished by increasing the width of each rear wheel by 1", and of each front wheel by 2", and by increasing the thickness of the wheel rims. The first 12 ton single was No.25307 for the Province of Palermo, Sicily, shipped in February, 1905.

When it came, however, to a roller of 15 tons nominal weight more drastic changes were required. It was decided that a compound was desirable, possibly for the ease of starting conferred by the simpling valve. The decision may also have been influenced by the fact that designs for compound versions of the 6, 7, and 8NHP traction engines were in the course of development in the drawing office, albeit in a somewhat leisurely fashion. Cylinders and engine work had already been designed for a 6NHP compound traction engine though no example of this was built until 1904 and these were conveniently to hand for the initial 15 ton roller in 1902. Other than the larger engine parts the extra weight was added by increasing the tyre thickness and wheel widths. The first made, No.23607, was sent to Henninger on January 31, 1902, whilst the second was No.23633 sold to W. & J. Scheid at Limburg, leaving on February 14, 1903.

Since there seemed to be interest in Germany in an even heavier version, design work was put in hand in the Autumn of 1901 for an 18 ton compound using the cylinder block designed but, up to that time, not used, for an 8NHP traction engine, i.e. 6" and 10½" in the bores and 12" stroke, mounted on a shortened version (4' $6^{11}/_{16}$" long compared with 5' $6^{1}/_{16}$") of the boiler meant for the same engine. This gave 6.06 square feet of grate area and some 130.60 square feet of heating surface, and the roller worked at 160p.s.i. The drawings produced for the 18-tonner came mainly from the hand of the master himself, namely Emerich Schmach, and were completed round about July, 1901. The drawings also included provision for a 20 ton (nominal) roller in which the only difference was in the width and thickness of the rolls. No 20-tonners were ever sold, despite being on offer for some twenty years. No.23677, the first 18 ton roller, was sent for exhibition on Henninger's stand at the 1902 German Royal Show held in May at Mannheim but no flow of orders resulted. Indeed, no other 18 tonner was sold until No.23729 left the Works at the end of April, 1903, bought by B. Ruthemeyer of Soest. This had followed No.23677 through the shops as a speculation and had been in stock since September 1, 1902. Ruthemeyer's order required his name to be cast on the slide valve cover with Garrett's name being only on the boiler number plate. Furthermore he wished to have it without forks, rolls, flywheel, steering chains and guards, saddle, funnel, axles, front scrapers, and firebars, all of which he intended to add in Germany. Ruthemeyer was in the process of launching himself into roller building and one supposes that he happened to find it inconvenient to build the engine and boiler for such a large roller in his own, or his German sub-contractor's, works. The project, however, must have been mutually satisfactory as a second example, No.24353, followed on June 6, 1903, minus the same parts.

Other 18 ton rollers sold to German buyers were No.24804 (April 11, 1904) to W. & J. Scheid, and No.27224 (June 11, 1908) for Henninger. The first had no wheels, forks, front scrapers, front axle or flywheel whilst Henninger's was without rolls but had both axles with it. These variations are an interesting example of the contortions through which a manufacturer had to be willing to put himself to secure an order. Selling rollers was proving to be an uphill task in a market place already mostly staked out by well-known and older established makes. Sales techniques abroad, though free of the limitations imposed in the United Kingdom by the price fixing ring enshrined in the Agricultural Engineers Association agreement, were open to competition that was always keen, often to the point of being cut-throat. The vast potential market presented by the Indian sub-continent proved to be a powerful bait. Burn & Co., as the principal agent for Garretts in India, wrote on July 17, 1902, that with competition in tenders as keen as it was the margin left for their own remuneration made it hardly worthwhile transacting the business. Could not Leiston, they asked, make a reduction in the prices charged to them? Replying on August 7, Garretts gave their agent what they called 'honourable discretion' to reduce the nett price quoted to them by a further £20 rather than to forfeit the order for a 10 ton roller. This, apparently, was all very well as far as it went, but Burns answer, dated August 21, asked what was to be done about an expected forthcoming tender for 8 ton rollers, expected to be keenly contested. In this case, in a further Garrett letter of September 11, they were authorised to use their discretion to the extent of £13. A week later they wrote a follow-up letter allowing £10 off the 'nett' price of a 6 tonner. Though these reductions may seem comically small in relation to today's monetary values they were substantial in the context of the time. The experience of Frank Garrett, Senior, in his years of trading with Burn & Co. had given him complete faith in their integrity in exercising these discretionary powers. The Garrett/Burn association subsequently had many successes in selling rollers to Indian customers although the profitability of some transactions was suspect. What is, perhaps, almost as interesting is the rapidity of this exchange through the post when all mail had to go by sea.

In order to give a boost to sales at home a 10 ton single, No.23642, was built in 'local show' finish for stock and put onto George Thurlow's stand at the Suffolk Show in June, 1902. George Thurlow (1837-1890) was Leiston born and had established himself at Stowmarket by the mid-1870s as an oil merchant, belt manufacturer, and dealer in and repairer of agricultural engines and machines. The firm he founded was carried on by his sons John and George (Junior), but in 1903, when John's health failed, it was incorporated as a limited company with George (Junior) as Managing Director. Relationships with the Garretts were always cordial. For their part the Garretts invariably placed their very considerable orders for belts with the Thurlows and, in turn, the Thurlows sold many Garrett machines long before the formal UK selling organisation was set up at Leiston in 1901 (see *Garrett Traction & Ploughing Engines)*. No buyer was forthcoming at the show but subsequently Thurlow brokered a deal with G.E. Gould of Ipswich, to whom the roller was sent on October 1, 1902. A short while afterwards Gould moved to Broadstairs, Kent, taking the roller with him, but he seems to have failed to prosper as it was back in Thurlow's hands and, a little later, at Leiston Works in the Autumn of 1905, probably repossessed under the hire purchase agreement. After reconditioning, the provision of boiler fittings to comply with German law, and the addition of a Bomford scarifier it was sold to Henninger on September 12, 1906. A further 10 ton roller sale arranged by Thurlow followed a similar pattern. No.23964, sent on October 6, 1902, to A. Bailey 'foreman driver to Norfolk County Council' at Wymondham was back at the Works in 1906 and that too was altered and sent to Henninger.

Also in 1902 the Works made a 10 ton convertible, No.23952, for stock. The definition of convertibles adopted in this series may call here for some enlargement. Convertibles came in two types. Firstly there were those which were primarily traction engines, capable of being converted to road rollers. These were longer in the boiler barrel than a roller designed as such, differed in the gears, and were fitted with differential gear and winding drums. Secondly there were others in which the primary function was rolling, with use as a traction engine as an occasional alternative. The first type are dealt with in *Garrett Traction & Ploughing Engines* and the second in this book. Though this may sound like an exercise in semantics such is not really the case. Garretts made the distinction in their books and I have followed their lead. The design of No.23952 illustrates this point. Although it had a winding drum and a perch bracket beneath the boiler barrel for use when dressed as a traction engine it was primarily a roller, not only dimensionally and in the gearing but also in the sense that it had no differential, and was fitted with a Bomford scarifier. No.23952 was completed in October, 1902, and appeared on the stock list of November 1. A month later it was sold to George B. Le Grys of Moat Farm, Heveningham, near Saxmundham, an old established threshing contractor. Like many in the same trade, he had realised that steam rolling for some of the rural district councils, who had responsibilities for all except main roads but often did not have the resources to purchase and maintain their own steam roller, might be a profitable adjunct to the seasonal round of threshing. How well it worked out with No.23952 we do not know but Le Grys later sold it to H.G. Rose of Beccles who last licensed it in 1926. Other convertibles were made and sold but the concept never really took a firm hold in the trade.

Although the firm otherwise had virtually no success in selling four-shaft heavy rollers directly to local authorities in mainland Britain, Belfast Corporation had Nos.24468 and 25384, both 15 ton compounds, delivered on August 21, 1903, and July 11, 1905, respectively, becoming Nos. 4 and 5 in the Belfast fleet. In part this lack of success in selling to Local Government arose from the long-standing amity felt by the senior director towards the Avelings, going back to the earliest days of traction engine building which made Frank Garrett, Senior, reluctant to canvass actively for home roller sales. Victor Garrett once recalled to me his father saying 'Avelings virtually invented the steam roller - they deserve the bulk of the sales', an attitude not entirely shared by his other sons and certainly not subscribed to by Victor. However soon after his father's restraining influence was removed the company, with his brother Frank in the chair, was committed to the AGE combine in which group policy reserved steam roller building for Aveling & Porter.

Once the design of the 10 ton single was established on its own type of boiler it underwent little development, such revisions as took place being mainly in refinement of detail. The most striking of these was the adoption of the swivelling fork and its concomitant type of forecarriage head first introduced with the three-shaft rollers derived from the tractor designs (see Chapter 5) but later applied to all rollers. The 10 ton four-shaft single had a circle of satisfied customers, both with German clients, particularly W. & J. Scheid who essentially bought only the engine, gearing, boiler, and tender, adding other components at their own works, doubtless varying the nominal weights to their own requirements in the process, and with Burn & Co, and their clientèle of Indian and Burmese municipalities. The reputation of Garrett rollers in India was very high. A copy is preserved in the Garrett records of a report, dated February 9, 1904, by W.B.McCabe, MICE, FIC., chief engineer of Calcutta Corporation. There is no indication of how Garretts came by the report or of the circumstances that prompted the writing of it. Possibly it was the engineer's retort to a suggestion of excessive partiality towards Garretts, but its unqualified approval of the rollers they supplied is a very telling testimony to their excellence:

> The reason for preference of Garrett rollers before Aveling & Porter's are that the former, besides being in all respects equal to the latter, contain in addition a number of modern improvements and advantages, all of which tend to safety, economy, and long life of the machines.
>
> Among the principal items in favour of the Garrett make are the following:-
>
> The shaftings and bearings are better fitted. Pinions are better proportioned, and the whole gearing is more accessible. Cylinder lubrication is continuous, a patent mechanical feeder being worked from the crosshead. Substantial lubricators being fitted to all bearings, several of which are dust-proof and oil-retaining.
>
> The boiler feed pump is fitted directly at the driver's hand and the double water gauges are brought right under his eyes. These gauges are provided with metal protectors and an automatic shut-off to prevent the escape of hot water and steam when a glass breaks.
>
> The combustion chamber of the boiler has a corrugated crown which, besides being stronger than a flat one, increases the heating surface. In connection with this a most essential fitting is provided, viz. a hand-hole in the boiler front at the level of the firebox crown, so that any accumulation of mud or other non-conducting matter is easily removed from this - the most vital part of the boiler, the plates being thus preserved.
>
> The self-adjusting scrapers for rear rollers are very effective.
>
> The workmanship and finish of the whole machine is superior, and the District Engineers assert that with the same stores much more work is done by the Garrett roller than by the Aveling & Porter.

One of the more unusual 10 ton rollers to be delivered in this period was No.25307 ordered by Grimaldi for the Province of Palermo. This was ordered as a 12-tonner and, notwithstanding the fact that a standard 10-tonner weighed in empty at over 12 tons, it was provided with ballast boxes to carry, including the weight of the boxes and brackets themselves, a further 4840 lbs. (2200kg). This was achieved by encircling the boiler, to the rear of the cylinder block, and the smokebox with two clamps, made out of 7/8" x 3"mild steel, in two halves, with lugs forged each end to enable them to be bolted together. To save the arrangement slipping round, four 3/8" holes were drilled through the upper part of each pair of bands and through the boiler barrel or smokebox into which block bolts were inserted and nutted on the outside. Each of the half clamps had a cantilever bracket of similar section material riveted on with countersunk rivets. Troughs - 15½" wide x 6" deep x 3' 3" long internally - made of 3/8" steel plate were carried on these cantilevers. Concrete blocks of the appropriate size and weight were made and added in Italy. Ungainly though the appearance of the finished roller must have been the weight distribution as between the front and rear rolls had been improved.

The first significant amendment to the concept of the 10 ton single was the designing of the so-called *Reading* roller in 1902, with Arthur Girling appointed to take the major rôle. The four-shaft arrangement was still employed but the cylinder stroke was reduced to 10" instead of 12", and the disc flywheel was reduced from 4' 2" diameter and 5" wide to 3' 3" diameter by 4" width. The wheel diameters were also reduced from 5' 6 ¾" to 5' 3" in the case of the rears and 4' 0" to 3' 8¾" at the front, the latter being changed, at the same time, from 2' 0½" to 1' 6 11/16" wide enabling the overall width to come down to no more than 6' 0". Coupled with other reductions in the capacity of the boiler and tender these design amendments allowed the weight in working order to drop from 13¼ to 10¾ tons. Why the new pattern of roller should have been dubbed the *Reading* at first glance seems obvious. The customer by whom it was ordered was Henry Lewis, the well-known owner of, and dealer in, engines, whose establishment was in Reading - hence, one might assume, the class name. However, when I was writing *Garretts of Leiston,* just on forty years ago, I was assured that Lewis ordered the roller thus to humour the preferences of the borough surveyor of Reading itself. Lewis could have used the roller on work for this gentleman so the story may have been true. If it was, the endeavour must have failed for there were no repeat orders. This first example for Lewis, carrying the Works No.24013 (February 19, 1903) was followed by a second (No.24014) for stock. This was dated June 1, 1903, and was shown at the Suffolk Show that month. The third of the batch, completed in December 1903, hung about until March 24, 1904, when it was sold as a 9 ton roller to Henninger, having had the boiler mountings amended to suit German law. Burn & Co. in Howrah bought occasional examples until 1920. From first to last eighteen *Readings* were made. No.24014, the second made, was not sold under its original number but went to W. & J. Scheid in the stripped down form required by Scheids as No.24805 in March, 1904, designated a 9-tonner! The *Readings* ought to have been, and probably were, more economical in operation than a standard 10 ton single, though no test results survive to establish the matter one way or the other.

The *Reading* design, however, came into its own as the basis of the 6 and 8 ton single cylinder rollers, code-named *Monarch* and *Ruler*, which weighed in empty at about 8 and 9 tons respectively. In these the cylinder was bored out only to 6½" diameter compared with 7" diameter on the 10-tonner, and a narrow flywheel - 2½" compared with 4" - was fitted. The bulk of the weight reduction was achieved in the width and thickness of the rolls. All of these 6 and 8-tonners went for export and accounted for almost 25% of all the rollers made.

The 15 and 18 ton compound rollers, never easy to sell, continued to be offered with decreasing success. The last 18-tonner to be built was No.32315 for the Cordoba Municipality in Argentina, despatched on July 10, 1914, only weeks before the outbreak of the first World War. The final 15-tonner, though sold as a 16-tonner, was No.30248 of February 12, 1912, for Gobo de Mendoza in Buenos Aires.

Towards the end of 1904 work began upon designing a compound version of the four-shaft 10 ton roller, intended, as with the heavier four-shaft compounds to work at 160 p.s.i. This used a boiler dimensionally the same as the 10 ton single which carried only 140 p.s.i. The code name given to this new design of rollers was '*Autocrat*'. The cylinders used were 5¼" and 8½" bore and 12" stroke, making the roller about 5NHP on the RASE rating. The remainder of the roller, second and third motion shafts, gearing, axles, wheels, and tender, were all as in the single cylinder version. The first to be sold was No.25424 for the Palermo (Sicily) office of Grimaldi, the firm responsible for a good deal of Garretts' trade with Italy. This was sent from Leiston on August 2, 1905. The second to leave was No.25789 in April, 1906, ordered via Henninger by Nickel & Seitz, who were contractors in Mannheim. This had the rollers added in Germany. It must have been successful for they had another (No.27564) in February, 1909. The 10 ton four-shaft compound had fair success until 1909 when the three-shaft compound 10-tonner based upon the tractor design became available. Thereafter sales went into decline though the odd example continued to be sold. In the course of its relatively successful period a batch of five, Nos.26755 to 26759 inclusive, was sold to the Sociedad Anglo-Espanola de Moto, delivered between December, 1907, and February, 1908, whilst No.27100 (April 18, 1908) was sold to Whitehead, Sumner Harker & Co. also for a Spanish client. Three (Nos.27175, 27870, and 27871) went via Mattos of Buenos Aires. No.27754 was sold by Grimaldi to the Ufficio Technico Provincale, Catania, via his Sicilian office in Palermo. No.28046, also sold through the agency of Grimaldi, had a cast steel swivel fork, thus showing the way to the future. For a number of years Garretts listed a compound version of the *Reading* roller, (given the code-name *CoReading*), utilising a boiler of the same dimensions as the *Reading* and the same general arrangement except for the cylinder block and crankshaft, but no example was sold.

Though the story of the four-shaft 'heavy' rollers is by no means over, it is a convenient point at which to leave it temporarily in order to introduce the beginnings of tractor building which was to become an influence on Garrett rollers. The idea of the '10 ton roller' that in reality weighed probably at least a third more was not dead in the home market but increasingly irrelevant to sales overseas where the real working weight came to be stipulated by enquirers.

Fig.10. *One of a pair of compound 15-tonners (Nos. 25380 and 26007) supplied to Mattos in Buenos Aires for the Ministerio Obras Publicas. These are the only 15-tonners known to have had extended awnings.*

Fig.11. Head-on view of a 12 ton compound roller, showing the older type of scrapers to the front rolls.

Fig. 12. (below) One of the series of 15 ton compound convertibles (probably No.26011, October, 1906) exported to Buenos Aires for the city of Cordoba, through the agency of M.E.Mattos.

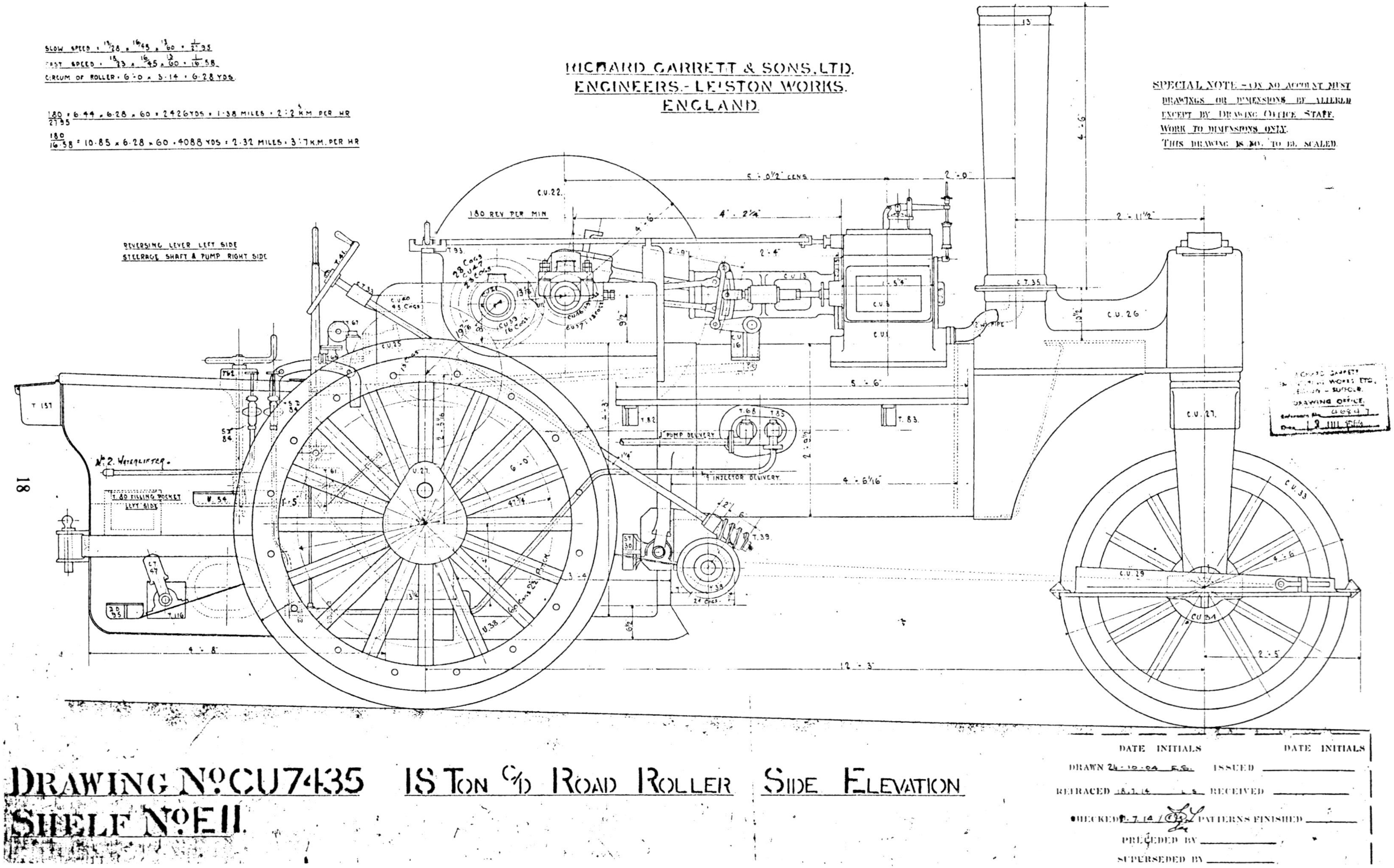

Fig.13. Gear-side elevational drawing No.CU7435, originally drawn by Emerich Schmach, of the 18 ton compound roller, after it had been retraced in July, 1914.

Chapter 2

Why Build Tractors?

In *Steam is the Essence* I gave an account of the vast vested interest that formed the opposition to attempts to supersede the horse as a prime mover in road transport. Notwithstanding the ramifications of this contra-force, by the mid-1890s a sufficient number of the influential had been converted to the cause of mechanical traction to secure the passage through Parliament of an Act to relieve steam and motor vehicles of some of the more onerous restrictions that theretofore had reposed upon them.

Apart from the additional freedom conferred upon motor vehicles generally by this legislation, the Locomotives on Highways Act, 1896, which came into force on Saturday, November 14, 1896, sanctioned the management by one man of mechanically propelled vehicles weighing not more than 3 tons. The promoters of the Act hoped that thereby not only would mechanical haulage of vans and drays, previously horse-drawn, be made legal but also economically viable. This change in the law encouraged the building of numerous designs of steam tractor most of which were not related to the general form of the traction engine, but one or two builders of traction engines applied themselves seriously to the problems of producing such a machine which, in effect, meant making a scale model of a traction engine, not much more than two-thirds full size but with parts sufficiently robust to withstand daily work in the hands of an ordinary driver, and capable of standing up to the abuse to which traction engines as a class were habitually subjected by both owners and drivers.

Of the number of traction engine makers who attempted the task the firm quickest off the mark and showing the maximum of dedication was Wallis & Steevens of Basingstoke. Others who tried were Aveling & Porter of Rochester, William Tasker & Sons of Andover, and Ruston & Proctor of Lincoln, the two former with greater success than the latter. Most of the traction engine firms, however, declined to embark upon what they deemed to be an impossible assignment. For a long while Frank Garrett, Senior, followed their example despite his sons having already talked him into agreeing to build traction engines and rollers, and a little later a steam lorry (see *Garrett Wagons - Part 1*). For their part the sons had perceived, or perhaps sensed, that the development of road transport made this a promising field into which to expand being, as yet, free of the intense competition and/or price fixing agreements that existed in the agricultural machine trade. The building of the first three Garrett undertype steam lorries proved to be a near disaster but the traction engines and rollers were beginning to show signs of success. Despite this Frank Garrett Senior remained sceptical about the logic of diverting the energies of the firm away from portable engines, fixed engines, and threshers, all of which were established money earners, into fields of manufacture where the prospects of reward were problematical.

Though the younger Garretts might have chafed under the reluctance of their father to agree to their plans to build a 3 ton tractor - which must, I am sure, have had some of its origin in his aversion to changes in the status quo - they might well also have reflected upon how little success in producing a really satisfactory machine *per se* or, more importantly, creating a market for those that were produced had attended the efforts of those firms who had already embarked upon the trade. Doubtless the senior director considered also that getting first an up-to-date design of traction engine and then a range of road rollers on to the market had already bespoken enough of the time of his chief draughtsman and drawing office staff, although the competence with which these two tasks had been tackled had probably reassured him to some extent. Despite this the poor performance of the first design of steam wagon must have reawakened his apprehensions. As Victor Garrett remembered the situation it was not so much that his father ever became a convert to the idea of building steam tractors as that his opposition to it slowly softened under the continual persuasion to which he was subjected.

In this atmosphere the development of a tractor design that would satisfy the conditions laid down by the 1896 Act came relatively low in the list of drawing office priorities. Possible design features were considered and sketches were made, but it was not until some six years after the Act had come into force that firm designs were put onto paper. In that interval much had happened to mechanical propulsion on the road. So far as the transport of goods was concerned the influence of the Self Propelled Traffic Association and its local branches, particularly that in Liverpool, had drawn to it men of influence, scientific ability, and experience of commerce. The members of the Liverpool branch included Alfred

Holt, the shipping magnate; S.B. Cottrell, engineer and manager of the Liverpool Overhead Railway; J.A. Crowther, borough engineer of Bootle; J.A. Norris, city engineer of Liverpool; and Professor Hele Shaw, professor of engineering at Victoria University, Liverpool. Whereas the Spurriers, father and son, of the Lancashire Steam Motor Co., and J.H. Toulmin and W. Norris, both of Coulthards of Preston, who were also members might have been considered to have personal axes to grind none of the first five could, by any criterion, be considered to fall into that category. The conclusion reached by the Association, as by most others in a position to form an opinion, was that the 3 ton weight limit set in 1896 was too low and that relaxation of this was required. Pressure upon the Government eventually secured the Heavy Motor Car Order, 1904, which granted an increase of 2 tons in the permitted weight, effective as from October 27, 1904, so that by the time the first Garrett 3 ton tractor, the story of which is detailed in Chapter 3, appeared on the market, the day of these very small tractors was virtually over.

Making larger tractors legal, however, removed the burden of doubt as to practicability that had deterred most of the traction engine makers from attempting a tractor. From a situation in which there were few who did there shortly arose one in which few did not. For some it turned out to be not much more than a dabble but for others tractors came to make an important contribution to their trading. Wallis & Steevens, who had been early in the field with their 3-tonners, soon joined the ranks of 5 ton makers but because of their addiction to low boiler pressures never enjoyed the marketing success their pioneering might have been thought to deserve. On the other hand, Aveling & Porter who, like Garretts, had produced a 3-tonner only with seeming hesitation, were to make a great success of their 5-tonners. Taskers, too, on their smaller scale of production made a success of tractors, as did Fosters. The other East Anglian firms all built tractors, but only Government orders in the 1914-18 war led to the completion of a reasonable output of them. Fowlers and McLarens each made good steam tractors but sold fewer, perhaps because of price, whilst Thomas Green, the third Leeds builder, made only a limited attempt.

The two younger Garretts' persistence in advocating tractor making must, therefore, be counted as evidence of acumen and sound commercial judgment, qualities not invariably displayed by either of them. In this case, however, they had chosen a good horse and were to be rewarded by winning races with it.

Fig.14. The gear-side elevation of a Garrett No.3 tractor in its original form.

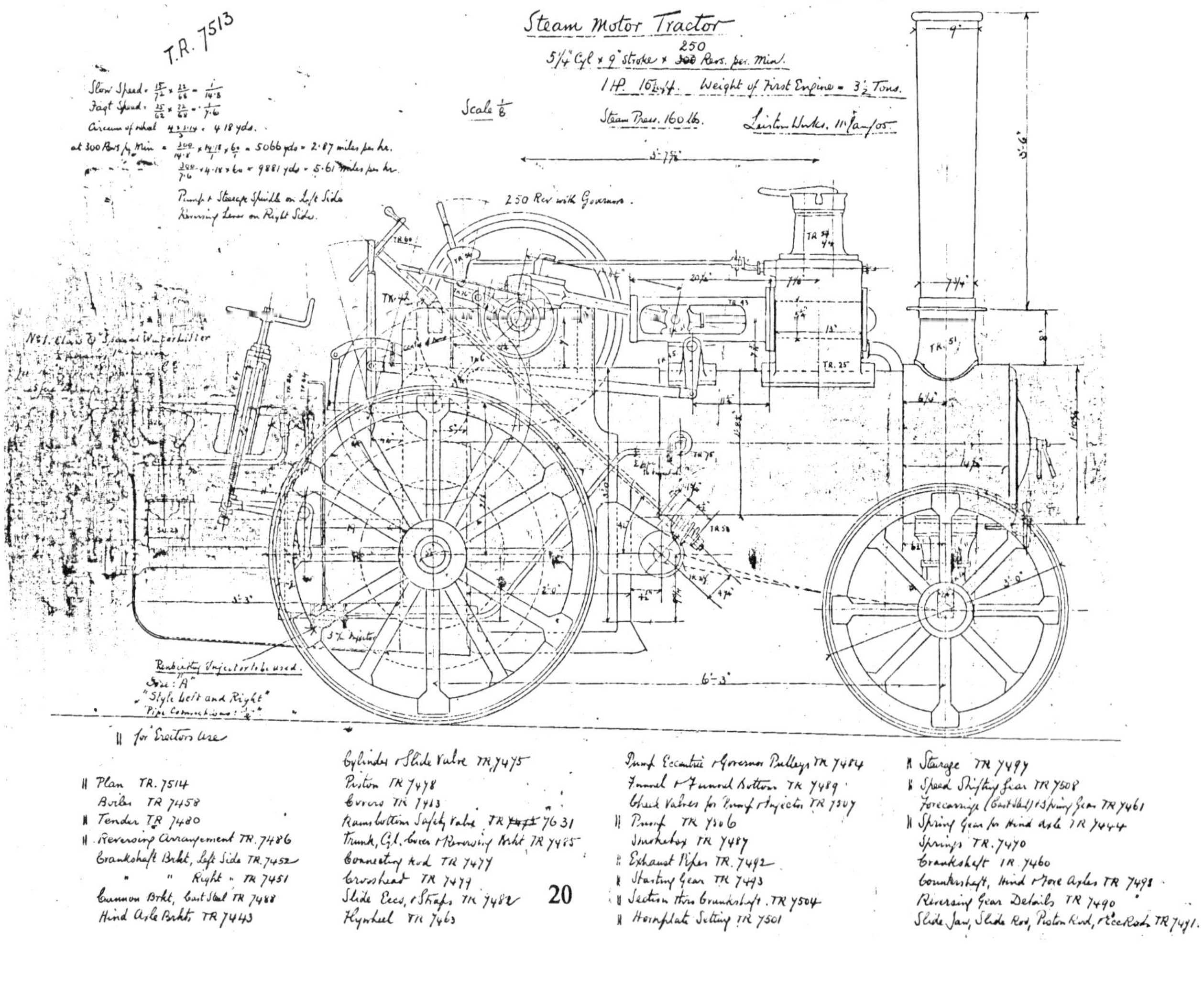

*

Fig.15. In front of the Customs House, Ipswich, against a background of barge riggings, No.3 tractor 25399 hauls 6 tons of barley for its owner, John Cooper (trading as R.J.Cole).*

***Fig.16.** The same tractor outside its owner's premises in Ipswich (between Commercial Rd., New Cardinal St. and Quadling St.) with a handsome furniture pantechnicon.*

***Fig.17.** No.25417 was fitted with the front tank to suit its first purchaser, Horace Bennett, who did not keep it long. It is seen here when owned by Albert French of Seaford.*
[The late W.G. French]

Fig.18. *No.25399 posed outside the* White Horse *at Leiston with one of the Works' own pole wagons and a load of deals.*

Fig.19. *The Carr brothers who were friends of the Garretts were also coal merchants in Leiston, although the coal trolley with which No.25399 is being demonstrated belonged to their Ipswich branch. Notwithstanding this trial they did not buy the tractor and remained faithful to horses.*

Fig.20. *A No.3 tractor, possibly No.25921, with the Garrett design of elm block rear wheels, posed outside the goods shed at Leiston station. No.25921 was ordered for Argentina with spring balance safety valves and an awning, and if this is it these must have been added after the photo was taken.*

Fig.21. *No.25565 was completed in November, 1905. It had governors, a winding drum, and a belly tank which, in turn, necessitated a longer smokebox to give clearance between the front wheels and the tank. The scene is the timber yard where Station Works was later built, and the roof is of Leiston Station.*

[A.J.Martin collection]

Fig.22. Garretts were keen to sell tractors for direct ploughing, and wider rear wheels and extension rings for the front wheels were fitted to a No.3 tractor for a demonstration. The picture is said to have been taken in late Summer, 1906 but the tractor cannot be identified beyond doubt.

Fig.23. Another posed picture with No.25399 and a 3' 6" thresher outside the Friends' Meeting House in Waterloo Avenue, Leiston.

Fig.24. The No.3 tractor was the basis of the 4 ton three-shaft roller No.31160 for Burn & Co, Calcutta, in Jan. 1913. [SRO]

Fig.25. The first No.4 single cylinder tractor, the standard design modified to suit the requirements of the Uganda Co. Ltd.

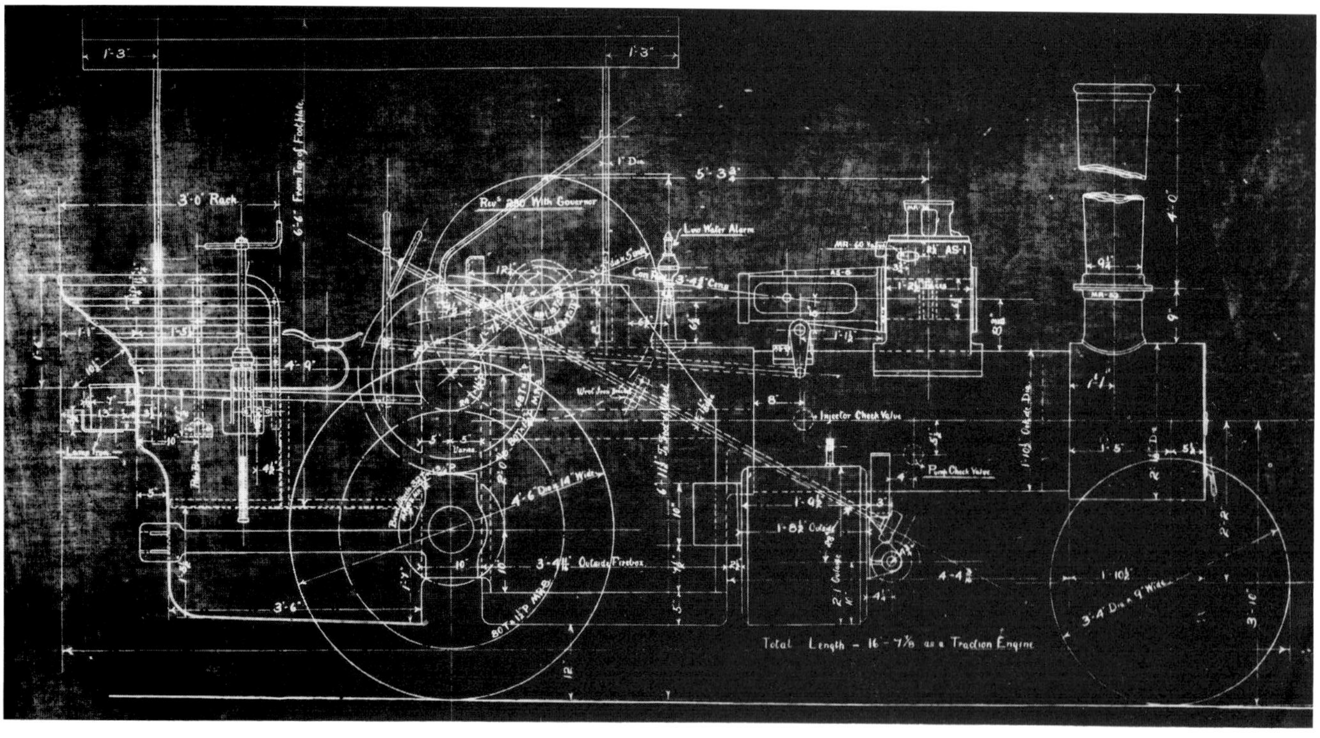

Chapter 3

The Three Ton Tractors

The task of making the drawings for the traction engines and rollers and, a little later, the several types of 4 ton steam tractors and their roller derivatives was divided between a group of four draughtsmen, Thomas Jones, W.H. Florey, Arthur Girling, and William Hewitt. On the other hand the bulk of the drawings for the 3 ton tractor seem to have been either Schmach's own work or to have been by F. Gordon Thomson, then relatively new in the drawing office and who later was much involved with the overtype wagons. It was Thomson who did many of the details though other draughtsmen made contributions. As was indicated at the end of Chapter 2 they were not done hastily, partly because the tractor still stood fairly low in the table of priorities in consequence of the senior partner's reservations about such tractors as a practical proposition. Doubtless, too, as with the launch of any new branch of manufacture a good deal of preliminary thought, coupled, as always, to an examination of how other firms had tackled the same problem, had necessarily to be gone through before actual drawing was commenced. By that time the concept of the 3 ton tractor as a scaled down version of the No.6 single cylinder traction engine was not novel, but Schmach was hoping to produce a more vigorous performer than had hitherto graced the market. The ever present problem, however, of how to do this without exceeding the prescribed weight limit proved to be probably the most onerous that had to be tackled. Indeed, it was never wholly solved. The finished tractor weighed 3¾ tons empty.

The final working drawings completed in January, 1905, provided for a 5½" diameter cylinder with a piston stroke of only 9", making it a 3NHP on the RASE rating. It was dubbed the No.3 tractor since Garretts never willingly used the term 'nominal horse-power' or the initials NHP, and abandoned such nomenclature altogether in 1904/05. The boiler barrel was 1' 8½" diameter and 3' 1½" long, containing 19 tubes. Wisely the designer had made them 1¾" diameter, refraining from committing the error, frequent at that period, of providing a much larger number of smaller diameter tubes, a practice which increased the heating surface but at the same time made it much less effective. Pressed to 160 p.s.i. and with a grate area of 2.41 square feet the boiler was a lively producer of steam *for its size*, but with only 42.80 square feet of heating surface it was to prove too small for much of its work. It was expected to generate enough steam to run the engine at 250 r.p.m. and produce a maximum of 18 brake horse-power - no small demand when it is realised that the 6NHP single produced 25 brake horse-power at its governed revolutions of 180 per minute. The disc flywheel was 2' 6" diameter and 4½" wide whilst the rear wheels, at 4' 0" diameter by a width of 8" were only 3" larger in diameter than the front wheels of the 6NHP and an inch narrower into the bargain. At 3' 0" diameter the 4" wide fore wheels were disproportionately large compared with the rears.

The gearing was cast steel throughout, as were the winding drum, cannon bracket, forecarriage, the axle boxes and their guides, and the fairlead brackets. With a width of only 1'10" and a length of 2' 2" the manstand was very constricted, such space as there was being further restricted by the presence of the pump and the reversing quadrant which combined took 6" off the width and by the brake cross shaft and the cannon bracket in the other direction. Only 19" separated the centre lines of the cannon and the brake shaft. A heavily built driver, like either Jack or George Newstead from the Test House, had just about room to get his feet onto it whilst still being able to reach the firehole.

As standard, the engines were fitted with a winding drum to carry 30 yards of ½" diameter wire rope, though a larger drum to carry 50 yards was supplied with some examples. The nearside rear wheel, running free on the axle, was secured by collar and cotter. The gear-side wheel, keyed to the axle against a shaft collar fixed with hexagon set screws, was held in position by a large washer plate and a 9" long hexagon set screw fixed into a tapped hole in the centre line of the axle, an inherently defective type of fastening. Though no accidents are known to have happened to a Garrett tractor from this cause a Tasker similarly equipped came to grief through loss of the set screw and hence the wheel. Springing of the rear axle followed Garrett traction engine practice using a transverse laminated spring and twin equalising levers so as to allow about ¼" of upward movement of the axle boxes in the guides.

A batch of three of these tractors was put in hand, the first of which was No.25399, and the erection of this began on May 19, 1905, in the experienced hands of erector charge-hand Walter Rogers. After it had been put through some preliminary trials in and around the Works, John Cooper of Ipswich arranged for it to go on practical trial with R.J. Cole, the removal and haulage contracting firm in Ipswich of which he was the owner. The Cooper family were long-standing acquaintances of the Garretts. Whenever the Works wanted to lay on a threshing demonstration, either to test a new design of engine or for publicity purposes it was to the Coopers of Knodishall Hall that they tended to turn for help. The business of R.J. Cole had been established about 1861, and by the early years of this century had premises bordered by Commercial Road, New Cardinal Street and Quadling Street. In its prime it was stated to have had upwards of 100 horses with appropriate carts and vans available but was, it seems, in decline when Cooper took control. He was in business close by in Commercial Road as a barley merchant. By 1909 the Cole name had disappeared. No.25399 was sent to Ipswich with Jack Newstead as its driver. It was fitted with a gong to give 'audible warning of approach', though one wonders what warning might have been needed over and above the sound of the steel wheels on the paving and the ringing of the cast steel gears.

The 4" wide front wheels were found to cut excessively into the road surface and a new pair 5" wide was sent to Ipswich on June 30, the drawing for future examples being amended accordingly. Coles liked the engine, and on June 6, 1905, agreed to buy it. By arrangement it was shown on the Garrett stand at both the Suffolk Show, held at Halesworth on June 8 and 9, and the Essex Show on June 14 and 15. Its career, however, was not an unqualified success. Two entries in the order book dated July 6 and 7 noted that a 'man to attend to tubes of boiler', that 'a governor to be fitted', and that 'a mudhole to be cut and a mud plug put in'. One imagines this might have been done by roughing out the hole with a ratchet brace and drill, cutting between the holes with a bull-nosed chisel and finishing with a file, but many fitters of that period were so adept at using a chisel that it might have been cut out all round that way. Wally Weston, a fitter with whom I worked on the Southern Railway, had begun his own career with the South Eastern in the 1880s under his uncle at Redhill shed. The first thing the older man did was to clamp a broken axle in a leg vice and tell Wally to cut through it with hammer and chisel saying 'When you've done it you'll be able to use a hammer and chisel'!

In September, 1905, No.25399 needed new piston rings and, six months later, what Garretts' ledger described as 'piston to be repaired'- which in reality probably meant a new piston. In the meantime, on December 6, 1905, a series of things had been put right. These included new keys and feathers to the main axle; two new axle bracket bushes; a new slide rod guide bush; new pump plunger complete with pin, cotter and brass; a new case hardened pin for the change speed gear; renewal of several bolts that had been broken or distorted in the hornplates; a new wire rope guide roller and repairs to the other roller and the mounting; two new pins in the starting gear; a new hose bracket; and an axle cap to replace one which had been lost. Whilst the latter and the hose bracket - perhaps, also, the damage to the guide rollers - might have been put down to heedlessness or misuse most of the work clearly related to early and excessive wear. The catalogue of replacements continued throughout 1906, beginning with a new steel casting for the forecarriage; two new pump valves and a new injector in May; new front springs (June 25); and front axle (July 9), and finishing with an 'overhaul and repair' on August 7. The year 1907 was rather better but still involved a new hind spring and 22-tooth pinion on July 23 followed by a new driving pin and new bevel pinion in November. Upon the closure of the Cole business in 1909 the tractor, after being briefly in the hands of a Mr. R.Tile of Ipswich, passed to E.A. Foley, the engine dealer of Bourne, Lincolnshire, from whom it went to Charles Clark of Cleethorpes and was lost thereafter to the Leiston Works records.

No.25417, the second in the batch, followed No.25399 through the shops and was taken into stock in June, 1905, only to encounter a pronounced lack of interest amongst potential buyers despite having been exhibited at the Royal Show at Park Royal from June 27 to 30, 1905. The sales department recorded it, over optimistically as it turned out, as sold to Horace Bennett of Stoke Ash, a farmer and threshing contractor, to whom it was sent on January 25, 1906. To increase the range of the tractor between stops and probably, at the same time, to reduce the tendency to be 'light on' i.e. rear up at starting, Bennett required it to be fitted with a 40 gallon rectangular tank carried on brackets bolted to the smokebox. The fact that the floor of the tank was 4' 4½" above the ground cannot have done much for the tractor's stability whilst as it was the full width of the engine the driver's view forward was very restricted. To enable the tank to be filled with water the water lifter was fitted on top of it and a connecting pipe was installed between this high level front tank and the tender tank. One hopes that a stop valve was put into this pipe but none is shown on the drawing. By February 13 it was suffering from leaking tubes as on that date the order book records that a tube expander was loaned to Bennett. Apparently this was not successful as four days later there was an order in the book for a boiler maker to go to Finningham Station to 'attend to tubes'. A further note in the same book records that on March 1 Bennett used the tractor to collect a second-hand engine from E.W. Cheney of Kelsale. Something seems to have befallen it on that task as on March 5 it needed two new lamps and the attention of a boilermaker at Framlingham 'to stop leaks'. Soon after this Bennett appears to have tired of it and it was returned to the Works where it remained for some while. It was next sold, in December, 1906, to W.G. Hole of East Dean, Sussex, who in turn sold it to Albert French of Seaford, to be used, among other purposes, for hauling a tar boiler on road tarring contracts. I never knew Albert French but his son, W.G. French, told me that although it worked quite well on those duties it was not otherwise very useful. After two further owners the tractor was last heard of in an auction at Hawkhurst, Kent, in 1924. The photograph of this tractor (Fig.17) shows how ungainly it must have looked.

The third member of this trio (No.25480) was completed in July, 1905, and shown at the Yorkshire Show held at Hull on August 9 and 10. On September 1 of that year it was sent on a week's trial, with Jack Newstead as driver, to the Jones Haulage Syndicate of Hampton-in-Arden. This somewhat peculiar organisation attracted much attention for a short while by extravagantly worded press releases giving details of its alleged national network of haulage routes and by placing large orders with various manufacturers for vehicles. Most of these came to nothing when Jones was exposed as a man having a vivid imagination but few assets beyond his glib tongue, and Garretts were but one of his victims. The following note concerning him appeared in *The Commercial Motor* of January 2, 1908:

> Two or three years ago, the Jones Haulage Syndicate appeared to loom very large on the horizon, and more than one firm proudly exhibited orders from it for fabulous numbers of steam wagons but we hear little or nothing of it now, and in a recent visit to Coventry I saw that Mr. Thomas Jones had a new occupation. The brass plate over his office indicated his connection with the Legion of Frontiersmen (Motor Section) Coventry Command. If this gentleman has hatched a new scheme which will result in the numberless steam wagons he ordered being brought into use I shall be extremely pleased; but------.

For its week with Jones the tractor was fitted with a supplementary 60 gallon circular water tank. After its return from this profitless episode it remained at the Works for some months, in use for demonstration to any likely buyer who presented himself. Early in April, 1906, it was let out on a month's hire to W.W. Howard Brothers Ltd., the London timber importers whose office was in Lloyds Avenue, E.C. but whose trade premises were at Commercial Wharf, Poplar. A charge of £5 per week was made for it, this to include 'an expert driver, oil, etc.' During this trial month it needed a new spring but, nevertheless, on April 30 Howards decided to buy it, and at about the same time they ordered a second similar tractor. During June No.25480 required new rear springs and the following month new ones at the front also. The rear axle was renewed in August, 1906, followed in October by a new right hand rear wheel, and in December by one for the left hand side. No record exists of any dispute with Howard Brothers over the cost of these renewals but they cancelled their order for the second tractor which is, I suppose, sufficient comment upon what they thought of the first.

Undeterred by the rocky ride that the initial three tractors were having, authority was given to proceed with a fourth, which eventually became No.25565 and was taken into stock during November, 1905. On January 20, 1906, it was sent off to Arthur Cook of Little Haywards Farm, Haywards Heath, Sussex. Cook requested the fitting of a winding drum, a belly tank, and Pickering type governors. The belly tank restricted the turning circle of the front wheels and necessitated a smokebox 21" long instead of the 14½" box originally fitted so that the front axle could be brought forward 10' 1/8" to give a wheelbase of 7' 1 1/8". This tractor, too, suffered rapid wear in much the same way as the first three, as before the end of April, 1907, it had required new piston rings, new pump plunger and gland, and a new slide valve spindle. In December, 1910, it was taken back in part exchange for a No.4 compound tractor and was on the Works stock for some time before being sold on July 21, 1911, to G.J. Deans at Baldock for £117 (nett), payable cash on delivery. Something seems to have gone amiss with this sale as the tractor was once again in stock at the Works in January, 1912, where it stayed on miscellaneous duties for some years. During this period it carried the registration BJ 2330, but after the AGE took over and reorganised the Works, it was sold to Goodman & Co. at Devonport. Later, in November, 1940, it was bought by H.A. Illingworth of Penrith to power a portable saw-bench, and from there it was sold in 1945 to Captain V. Hewitt at Cemlyn, Anglesey, thereby surviving into preservation. It now leads an extraordinarily pampered life in Holland.

One of these first four engines spent some weeks on hire to the North Eastern Railway from December 15, 1905 to January 15, 1906. The formal order from the railway was dated November 29, 1905, and the account for the hire was sent in February, 1906. It was registered as BT210 during the time the railway company had it, but it cannot be identified with absolute certainty though there are strong pointers. Firstly No.25399 may be eliminated because at the relevant times it was working for Cole in Ipswich. No.25417 might have been a candidate as it was in Garretts' possession at the right times, although the interval of only ten days between the ending of the hire period to the North Eastern Railway and its departure to Horace Bennett seems tight. The schedule for No.25565 is even tighter. Although it was finished during November, 1905, it spent the days of December 4-8 on display at the Smithfield Show, doubtless with a couple of days added each end for setting up and removing, whilst the five days (January 16-20) that would have been available between finishing at Gateshead and arriving at Haywards Heath, refurbished and repainted, seems impossibly short. Thus No.25480, which was known to have been serving as a demonstrator during the period in question, remains as the most probable candidate.

Still seeking the elusive crock of gold the firm built yet another example (No.25729), again with a belly tank and a warning gong. In March, 1906, in 'exhibition finish' it was sent to Grimaldi for display on his stand at the Milan Exhibition. Sadly it found no buyer and was back at the Works in time to figure in the monthly stock summary of March 1, 1907. It was given the new number 26580 and took some part in the 1907 show circuit in the course of which it visited the Wiltshire Show at Devizes on May 22 and 23, the Royal Counties at Maidenhead from May 28-31, and the Bath & West at Newport from June 5-10. A Pickering governor was added and an awning over the tender. In this form it was sent to agents, E. John Smith & Co. at Cowbridge, Glamorgan, but once again the homing instinct was too strong and in January, 1908, it was back at the Works, only to be sent to Grimaldi for the second time on February 18, 1908. On this occasion it did not return.

This was followed by another pair, Nos.25781 and 25794, in stock by March, 1906. Turned out in Royal Show finish and adorned with a copper-topped chimney, No.25781 was shown at the Automobile Exhibition in the Royal Agricultural Hall, Islington, from March 24-31, going on later to the Devon Show at Tavistock on May 22 and 23, the Bath & West at Swindon (May 31 to June 5), the Royal Counties, Portsmouth (June 12-15), the Norfolk (June 20-21), and the Royal at Derby (June 27-30). This, too, had a belly tank and a warning gong and also, for a reason not now clear, a rod for working the blast cock from the footplate. It was on the books for longer than most continuing on the demonstration circuit until September, 1908, when it was given the new number 27242 and sold to Grimaldi in Italy. The second of the pair, No.25794, had a less nomadic life, being sold on May 9, 1906, to Parker Brothers (Mildenhall) Ltd. The first eighteen months in their ownership were fairly turbulent. In January, 1907, it needed new rear springs and a main axle bush, followed in June by a complete new forecarriage and turnplate, surely to remedy the effects of an accident. Whilst this was being put on the order was given for the tractor to be thoroughly overhauled and repaired and, significantly, for the 'clutch gear to be altered to prevent both gears working in at the same time'! Either the engine gobbled up driving pins or the driver was careless about the way he put them in as it had two new pins on October 5, 1907, another single pin on October 14, and a further two on December 20. Concurrently with them it had two new countershaft bushes, a main shaft bush, a new 22-tooth pinion, and a new slide rod. After this things settled down and spares were still being supplied in 1923, so that it had a reasonable career at Mildenhall.

A further tractor followed this pair, almost neck and neck, through the erecting shop. No.25921 was finished in late May, 1906, with wood block hind wheels and an awning over the tender, attending the 1906 Suffolk Show at Ipswich (June 7 and 8), and the Essex Show (June 13 and 14) before being sent on June 27 to Mattos in Buenos Aires for a customer named Alberto Casares. How it fared is not recorded. A similar paucity of information accompanies No.26719, shipped on November 21, 1907, to John Chambers & Sons Ltd. at Auckland, New Zealand, via Keep Bros.

Three more tractors, all for Grimaldi, were built in 1908 as Nos.27237, 27242, and 27274, exported in September and October, 1908. No particular comment is called for except to say that while the first two had no awnings the other had an awning over the tender and manstand, and also had a flywheel brake. The third tractor disappeared from Leiston view on reaching Italy. No.27237 remained in Grimaldi's hands, used only from time to time as a demonstrator, for a period of four years. As he was not obliged to pay for an engine until it was sold, Garretts eventually recalled it so that it was back in Works stock by the end of December, 1912. It remained at Leiston, used from time to time on Works duties, until the opening months of the 1914-18 war when it was sold to Southern Transport at Brighton as their No.7. After a repaint in the customer's colours of buff with red wheels, and the fitting of Ramsbottom safety valves, and a Vacuum Oil Company 'Pony' size lubricator, it left Leiston finally on November 24, 1914, with the registration BJ 2311. When Southern Transport had finished with it at the end of the war it was sold to Ernest West of Burgess Hill who converted it to a roller. Later he sold it to F.J. Dunkerton Ltd, the quarry owners and road and drive contractors of Horsham, Sussex, who in turn traded it in to Wallis & Steevens Ltd. in part exchange for a motor roller. It was then sold to Ronald Weller of Cranleigh, Surrey, and later still passed to W.H. Hooker of Langley, Kent, before coming into the hands of Chris Lambert of Horsmonden, Kent. Though a founder of the preservation movement Chris was not moved to save this particular roller and in 1952 it ended up as scrap with J.W. Hardwick & Sons of Ewell, Surrey. Whilst the name of the purchaser of No.27242 is unknown the tractor itself had a reasonable length of life. In 1929 an order for a new main drive pinion was received from Tremontani & Figli, then acting as Garrett agents in Italy as Grimaldi had become insolvent.

By the end of 1908 the emergence of the No.4 tractors as the better selling proposition had led the company's interest in the No.3 to wane. Yet after a lapse of over two years and just as sales of the No.3 tractor were thought to be completely dead an order was received from Pilter's Dieppe office for a No.3 in Royal Show finish with an extended smokebox and dynamo stand, geared pump, Pickering governors, and an Empire oil pump. Could this, one is tempted to wonder, have been for a French showman? It was sent off on May 25, 1911, as No.29013. Almost within months Pilter ordered another No.3, this time from his Marseille office. The tractor was to be fitted with a Continental style open tender, wire top to the chimney, and both a buzzer and a gong. Unfortunately I have been unable to find a photograph of this interesting machine which became No.30140, despatched on March 27, 1912.

Even after this one more was sold. This was No.33215, sent on September 17, 1918, to the Castlemahon Co-operative Dairy Society Ltd. at Castlemahon, County Limerick. When in Ireland in 1965 I spent some time tracing what became of the engine. The creamery was still running and the staff were helpful. It was recalled that it had been sold to a machinery dealer opposite whom they promptly telephoned. In turn, he recollected that he had sold it to a farmer named Paddy Sullivan at Feenagh, ten miles or so away. With some difficulty we found the Sullivan farmstead. Mr. Sullivan was out but Mrs. Sullivan vouchsafed the information that he had kept it in the shed down the field for years, but in the end had been offered by a scrap dealer a price he could not afford to refuse - no less than £15! Incidentally, this tractor was one of the few to have an awning, in this case extended over the motion work.

Had this rather meagre output of No.3 tractors had to stand on its own economically, it would have presented a poor picture, but fortunately the tractor design later formed the basis of a series of seven steam rollers of 4 ton nominal weight which redressed the situation to some extent. All were for export, and four were on *Colonial* boilers with larger fireboxes. The first of these, and also the first with a Colonial firebox, was No.30268 (June 1, 1912) supplied to Mattos of Buenos Aires for onward sale to the Ministerio Obras Publicas in Argentina. The weight of the roller in working order was about 4¾ tons - much closer to the nominal weight than was usually the case. Cast steel forks were used and the rolls were grooved 2" deep, as can be seen in Fig.137, by the expedient of having rings shrunk on and studded to the parent plate of the roll rim. The rolls themselves were of plated, as opposed to spoked, construction and could be water ballasted if required. The steering bobbins were on the ends of the cross shaft outside the brackets, and the design of forecarriage head was that used on all three-shaft rollers as described in Chapter 5. There was an awning over the manstand, probably as much for protection against strong sunlight as against rain. Curiously, in an engine presumably, from the use of a Colonial firebox, intended to burn wood fuel, only a very limited space was allocated to carry it. No record could be found at the Works of the reasons for the grooved rolls though consolidation of earthwork dams or embankments has been suggested. Two further and similar examples (Nos.30301 and 30302) with plain rolls but each with supplementary water tanks 3' 6" x 2' 6" x 1' 6" slung either side of the tender were supplied in March, 1912, to E. Weber of Brussels for service in the Belgian Congo, lettered No.1 and No.2 respectively. Three were sold in 1913, two to Burns in Howrah, and one to Henninger, after which there was a long time lapse until No.34960, the final example, was sent to Jessop & Co. in Rangoon on February 25, 1927, fitted with an 'extended' (i.e. full length) awning.

Fig.26. *No.26432 (March 28, 1907), a No.4 single cylinder tractor exported to J.Macalister in Invercargill, New Zealand. It is believed to be the sole remaining No.4 single.*
[T.J.M.Keenan]

Fig.27. *The first No.4CD tractor to be invoiced to a client was No.26063, bought by Jacob Studt in May, 1907, after he had had it on trial for several weeks. No dynamo bracket was fitted until 1920 when it was owned by Abraham Coneley.*

Fig.28. *The North Eastern Railway were the first to order a 4CD, but their lengthy inspection procedures meant that it was the second to be delivered. In the picture, said to be in York, the Glover wagon with which it was sold carries a test load of railway brake blocks. The tractor, No.26479, was one of the few delivered without an awning.*
[A.J.Martin collection]

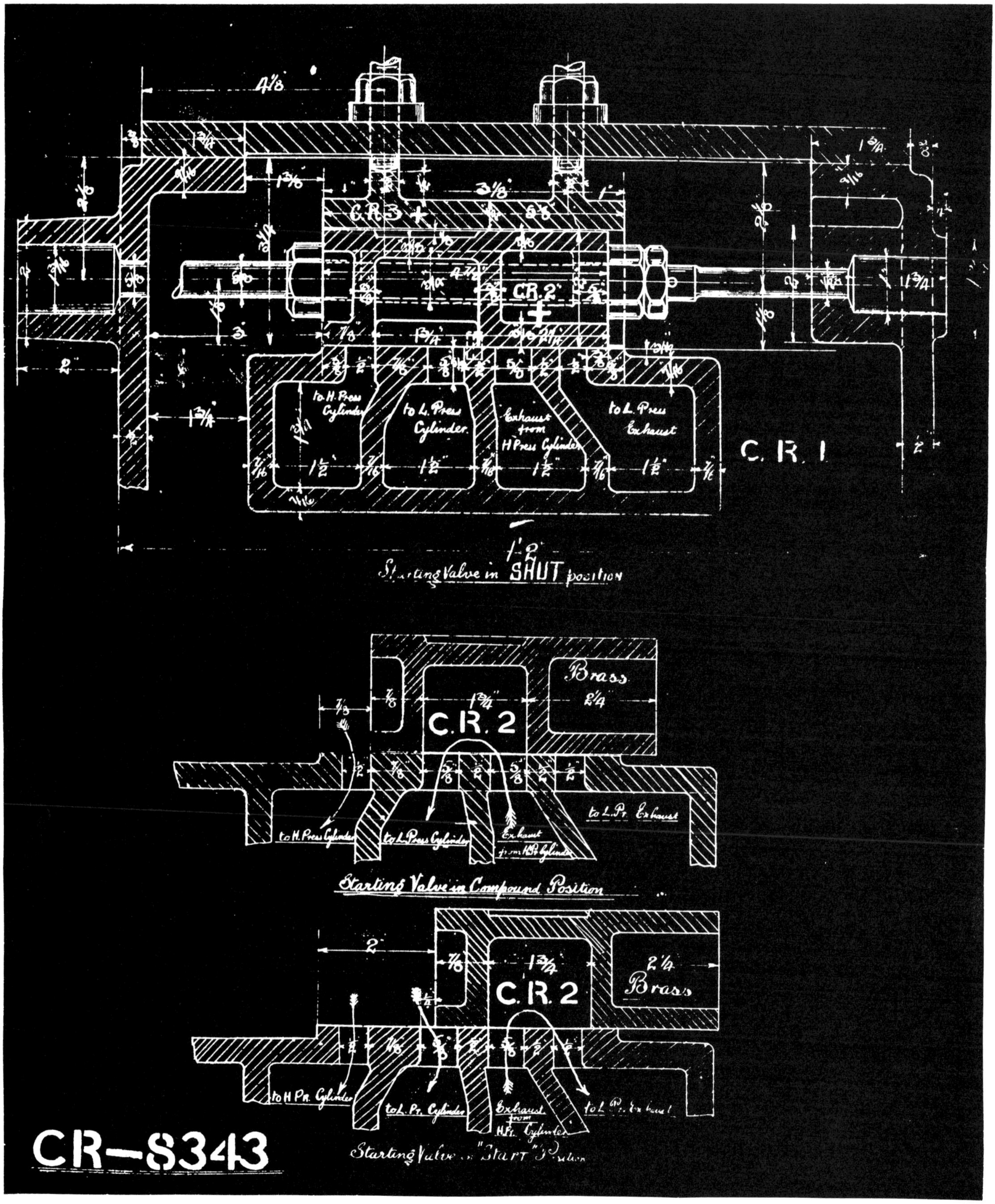

Fig.29. Drawing CR 8343, showing how it was possible to run a tractor on 'double-high' for extended periods.

Fig.30. *No.27884 (October 28,1909), the second of the two 4CD tractors sold to John Studt, in the Paint Shop (later No.9 Parts Store) adjacent to Haylings Road, Leiston. Note the shallower pattern of belly tank and the tubular water heater upon it - one of the first fitted. The man on the right is John Creasey, the shop foreman.*

Fig.31. *Numerically the next tractor was No.27886, seen here ahead of No.30050 when both were owned by Noah Etheridge of Blythburgh. No.27886 went to S.P.Robinson of Stanton, Suffolk, in 1913.*
[A.J.Martin collection]

Fig.32. *No.27886 was in Robinson's ownership when it was photographed passing under the Ipswich to Bury railway near Thurston.*
[K.Leighton collection]

*

Fig.33. A picture from 1914 showing Sidney Robinson posed in front of No.27886.* [A.J.Martin collection]

***Fig.34.** No.28745 was sold new to A.G.Cook of Haywards Heath, Sussex, to replace his No.3 tractor No.25565 which was taken in part exchange. About 1927 it was sold to Tom Smith, the Shoreham, Sussex, showman and in 1934 went to Leslie Searle of Parsonage Farm, Horsham, in whose hands it was when photographed in 1938. The awning is of later date than the engine and is not of Garrett origin.* [A.J.Martin collection]

***Fig.35.** E.W.Rudd had tractor 27088 in May, 1908, primarily as a demonstrator - he was a Garrett agent - with wood block rear wheels for running on London streets, but this is probably it, equipped for the time being with straked wheels whilst it was being tested at gun haulage.*

Fig.36. *The first two superheated tractors (Nos.28068 and 28076) were completed on March 1 and April 2, 1910, respectively. The first went to Argentina and was not heard of again, but No.28076 (above) went to the 1910 Brussels Exhibition, afterwards coming back into stock until sold early in January, 1911, to L.P.Stokes of Down Farm, West Kington, near Chippenham, who used it for direct ploughing fitted with 12" wide rear wheels having 6" bolted extensions and 20 spudlets, and 9" wide front wheels with detachable angles. He also had the drawbar attachment at two levels, the lower for pulling the implements and the upper for haulage.*

Fig.37. *No.29771 (September 7, 1911) was new to F.Matravers & Son of Coggeshall, Essex. In 1924 it was bought by William Barker, Junior, of Ipswich, and is seen here in his ownership at Christchurch Park, Ipswich, in 1935.* [R.G.Pratt]

Fig. 38. *This is a well-known picture but worth repeating. It shows tractor No.31025,* East Lynne, delivered *on November 28, 1912, to Thomas Cottrell, whose permanent address was at Droitwich but who was then, as he often was, built up with his bioscope show at Leiston. Tommy is standing by the rear wheel and the driver is George Newstead, from Garretts' test house. According to the late Frank Waddell, whose father took the photo, the fit-up (in the background) was then built up in Valley Road, Leiston, and the tractor was actually in use.*

Fig.39. *Tractor No.31118 (February 14, 1913) spent nine years with its first owners, the Devon Trading Co. Ltd., before being part exchanged for a wagon. Garretts sold it to Reynolds, the engine dealers of Bedford, who, in turn, placed it with W.Price of Walford, Herefordshire, by whom it was owned when photographed soon after purchase, hauling a very large ash butt.* [A.J. Martin collection]

Chapter 4

The Five Ton Tractors 1906-1911

Very shortly after the design of the No.3 tractors had been started in the drawing office work began on drawings for a four nominal horse-power version. Indeed for part of the time work on the two designs proceeded concurrently. By the late Summer of 1905 the making of the drawings for the No.4 single were well in progress and they were complete by early Autumn - the plan (drawing MR7836) was finished and dated on October 25 with the cross section (MR7839) following on November 2. With the advantage of hindsight knowledge of how the trade in compound steam tractors ultimately was to develop the temptation arises to question why time was wasted in designing single cylinder tractors at all. The fact was, however, that experience up to that time suggested that single cylindered engines were most likely to find popularity. Single cylindered traction engines and rollers outsold compounds by a considerable margin, not only with Garretts but with most other makers as well. Whilst, no doubt, this factor weighed as much at Leiston as elsewhere, there was the further consideration that Garretts' overseas customers were showing an interest in much lighter types of traction engine than had been the case thereto, giving rise to the possibility of using components of the tractor as a basis for such a machine. As narrated in *Garrett Traction & Ploughing Engines* this led to the building of 4 and 5NHP unsprung traction engines, but at first the main thrust was toward a spring mounted tractor for one-man operation in Great Britain.

The cylinder was cast from pattern AS1, made for the *Reading* type four-shaft roller, bored to 6½" diameter by, of course, 10" stroke; the trunk guide to the crosshead (AS8) came from the same source as did the motion bracket (AS91). The steering was borrowed from the No.3 tractor, together with the filler pocket, the pump body, and sundry flanges and connectors. The governor bracket came from the 6NHP traction engine, and the hose bracket (to pattern L120) from the ill-fated first undertype wagon. All other components were of new design.

As in the No.3, Stephenson link motion reversing gear was used. The webbed and balanced crankshaft was 2¾" diameter, and the offside end was squared to receive the change speed gears, which were shrouded cast steel, 2$\frac{5}{16}$" wide overall (with 2" wide teeth). At the flywheel end, inside the hornplate, was the pump eccentric. The solid flywheel was 5" wide on the face and 3' 3" in diameter. The 3¼" diameter second motion shaft ran in bronze bushes in the cannon bracket, the 24 tooth driving pinion meshing with the 80 tooth driven ring (MR8). These gears were 2½"wide. The ring was bolted to the differential centre plate (MR7) running free on the axle, with the two differential bevel pinions engaging with the two bevel wheels, one of which (MR12) was keyed to the axle, the other (MR10) being free on it and driving the nearside wheel by one or two pins, as ordered for individual tractors. The differential could be locked by inserting a longer pin to engage with one of the four holes in the centre plate. At the other end of the 4" main axle the collar MR25 was keyed to the axle with a single pin drive to the road wheel. This 4" axle proved to be a potential weakness and in September, 1908, was redrawn on drawing CR9776 as 4$\frac{1}{8}$" nickel steel. The cannon bracket was a 'bought-out' steel casting until 1918 when the wartime need for the maximum number of castings to be made at home led to the substitution of cast iron. This gearing was modified, and the ratios reduced, at the time when the driving wheel diameter was increased from 4' 6" to 5' 0". Casting MR8 became MR86 with the number of teeth increased from 80 to 86 whilst the mating pinion was decreased from 24 teeth to 18, all still at 1½" pitch. The crankshaft and intermediate shaft pinions, whilst remaining at 1¼" pitch, were dropped from 30 and 68 to 28 and 70 respectively.

The springing of the rear axle was provided by a single seven-leaf transverse spring, 2' 0" long, coupled to the equalising lever by an 1¾" diameter eye bolt, passing through the buckle and leaves of the spring and adjustable below by a hexagon nut secured by a collar and split pin. When correctly adjusted the axle box had $\frac{1}{16}$" clearance at the bottom of the guide and $\frac{5}{16}$" at the top so that the cushioning effect of the spring could fairly be described as small, if not minimal. The driven wheels were 4' 6" diameter and 10" in width, the fronts being 3' 4" diameter and 6" wide. Each rear wheel had a 2½" wide brake ring riveted on. The brake blocks were hung on a 2" diameter shaft which crossed the floor of the manstand and was actuated, via a crank, by a vertical shaft and quick thread, bracketed onto the offside of the tender. The tractor was arranged for left-hand steering, with the bobbin mounted on the front of the belly tank which was hung under the boiler 2½" forward of the firebox outer plate and extending downwards to the level of the ashpan bottom, thus necessitating a damper opening toward the offside.

The boiler for the No.4 tractor, designed, as would be expected, around the Garrett sine curved stayless firebox crown, used twenty-two 1¾" external diameter x 11 gauge ordinary tubes and six 1¾" stay tubes with 3/16" wall besides which there were two 1 3/8" diameter longitudinal rod stays between the boiler front and the smokebox tubeplate. These had a forged eye at the firebox end, fixed to the plate with a 1 1/8" pin and two 3" x 2½" x ½" angle lugs riveted on. At the other end these stay bars were threaded and secured with hexagon nuts, washers, and back nuts. The plate thicknesses were: tubeplates 9/16"; firebox crown and hornplates ½"; other inner firebox plates 3/8"; front plate 7/16"; barrel 3/8". The boilers were thus very strong, and in use proved to be exceedingly good generators of steam, but, at the same time, were probably as low in weight as could have been expected. At a working pressure of 160 p.s.i. at 230 r.p.m. the tractor was capable of producing 25 BHP as a maximum. This compared well with the 18 BHP produced by the No.3 at 250 r.p.m., using about 25% more coal and water per brake horsepower hour. The weight of the tractor, empty, was stated to have been 4.85 tons, and though this statement may prompt one to reach for a grain of salt the design did make a very serious attempt to meet the requirements of the regulations.

There was an element of irony in the fact that after the design had been worked out in this way to comply with the regulatory requirements of the home market the first order to be secured should be for an engine for overseas use, obtained, in all probability, through Frank Garrett Junior's contacts among the purchasing engineers and confirming houses in the City of London. William Millar & Co. of Fenchurch St. required a light and fairly fast road locomotive for their clients, The Uganda Company Ltd. To get the order, however, a number of modifications to the, as yet unused, standard design were necessary. It was required that a colonial firebox should be fitted; that the rear wheels be increased in width from 10" to 14" and the fronts from 6" to 9"; that the capacity of the belly tank be increased by 25 gallons; and that the tractor should be convertible to a road roller. Other stipulations were the fitting of a low water alarm, two Klinger gauge glasses, and the provision of a wood-rack. In fact, the differences were sufficient for a new elevation and plan to be called for (MR7904) together with no less than twenty-one separate component drawings, some admittedly for relatively minor items. Virtually all these drawings were completed by the first week in December, 1905, notwithstanding the fact that the order had only been placed on November 29, 1905, and the finished engine, No.25646, was ready for despatch on January 23, 1906, which suggests that work might have been put in hand ahead of the order.

The next No.4 single cylinder tractor made was No.25737 for Bunger & Leyrer in Germany. This was much nearer to standard apart from having a spark arrester, and a system of rods to enable the blast cock to be worked from the footplate. It left on February 23, 1906, but was back on the works in January, 1907. Having been stripped of its Continental features, in April it was sold as No.26515, via the agency of E. John Smith of Cowbridge, to W. & T. Jenkins Ltd. of Brynsadler Brewery, Glamorgan. No.25957, built for exhibition at the 1906 Royal Show at Derby, was sold by the same agent in November, 1906, to L. Aston of Dinas. This and subsequent No.4 tractors had an exhaust steam superheater in the smokebox.

Concurrently with these developments others had been taking place which were to redefine the future course of tractor building at Leiston and to reduce the already scanty sales of the No.4 single still further. After No.25957 only fourteen were sold, all of which, with the exception of No.26515, were exported. Macalister in New Zealand was responsible for selling ten, nine of them with Colonial fireboxes, and none with any remarkable special characteristics. The last to go was No.31126 - the final No.4 single made - shipped in 1912. Of these No.26432 (1907) still exists in New Zealand. Burn had No.26621 for India; Waygoods in Sydney, New South Wales, had No.26008, whilst No.26041 was sold to Gilbert Hamer & Co. of Durban. The cylinder block and other engine components and the boiler of the No.4 single cylinder tractor formed the springboard for the series of light traction engines built for export, mainly to the present Baltic States.

It was recounted in Chapter 3 how the North Eastern Railway had had a No.3 tractor on trial but by the end of January, 1906, had returned it as too small. The fact that George Gibb (later Sir George), General Manager of the North Eastern Railway from 1891 to the end of 1905, was married to one of the Aldeburgh Garretts might be seen as an explanation of how the railway's enquiry for a tractor reached her Leiston cousins rather than going to a company such as John Fowler & Co. that was one of its customers.

It may indeed be the explanation but it must be remembered that the purchase of such a relatively trivial item would not normally have fallen within the purview of a General Manager. Alternatively the Garrett tractor, No.25480, at the Hull Show could equally well have been what attracted the attention of the railway's officers.

Discussions between Leiston representatives and the staff of the Chief Mechanical Engineer's office of the North Eastern at Gateshead led to the latter being shown the drawings of the No.4 single cylinder tractor. I rely upon the recollections of Victor Garrett, admittedly quoting hearsay from his brother Frank, when saying that the NER people expressed a preference for a tractor that might be a little quieter in the exhaust and rather easier to start than a single. Given the interest both at Gateshead and at Leiston in the use of compounding this led on naturally to the propounding of a No.4 compound tractor equipped with a simpling valve.

As with the No.3 and the No.4 singles Emerich Schmach himself took a leading part in the additional design work, with assistance from Gordon Thomson. For example CR8342, the drawing for the compound cylinder, was Schmach's work. How far he had formulated proposals for the relatively few additional drawings needed for the compound ahead of the discussions with the NER cannot now be determined. The same design of boiler was used but pressed to 180 p.s.i. rather than the 160 p.s.i. of the singles. The cylinder, crosshead guides and crankshaft assembly drawings were all, of course, new as were the general arrangements. Most drawings were done by September and October, 1906, the drawing (CR8405) showing the offside elevation and centre lines, for instance, being dated October 25.

In the first cylinder design on drawing CR8342 the diameters were 6⅞" and 4½" by 10" stroke. These diameters were reduced very soon to 6¾" and 4⅜", giving the same power output as the singles, although when the first of the range of three-shaft rollers based upon the tractor were in the making (see Chapter 5) the cylinders were bored out to 6⅞" and 4½". The cylinder block drawing was subsequently renewed as CR11018 but the bores for the tractors were again altered back to 6⅞" and 4½" by a note dated August 5, 1908, and initialled by Thomas Jones. Tests conducted on the first two examples to be built were to show a marked economy of fuel and water. Whilst the coal consumption of a No.3 tractor per brake horse-power hour was of the order of 5lbs, and in the No.4 single something like 4lbs, with the No.4 compound, referred to hereinafter as the 4CD, the figure was further reduced to about 3¼lbs. The corresponding figures for water were 39, 32, and 26 lbs respectively. Though the tests on which these figures were based were of necessity limited in number, and possibly conducted with less than scientific precision, the downward trend was clear.

As the talks with the NER plodded along and the railway took its time in considering the drawings for the compound, the decision was taken to build an example, subsequently numbered 26063, for exhibition instead of a No.4 single at the 1906 Smithfield Show at the Agricultural Hall, Islington, from December 10 to 14. The work of erection began on November 11 with Tom Staulkey in charge, a reflection of the confidence his employers had in him. Meanwhile the North Eastern Railway negotiations rumbled on, culminating in agreement by Garretts to build the tractor and allow it to be tested in service before acceptance. The work of erecting this tractor, No.26479, was allocated on February 13, 1907, to Walter Rogers, another very trusted erector, though a little younger than Tom Staulkey. After being taken into stock in April it seems it was sent to Gateshead for trials. The actual trial dates are lacking but spares were sent to it there during May, 1907. The trial period appears to have been satisfactory as the railway finally purchased the tractor, the date of the formal order being July 19, 1907. Whilst I have no details of its whole career with the railway company, in a letter to Garretts from T.W. Carr of the North Eastern's motor vehicle department at York he stated that initially, at least, it was used at Northallerton for 'dealing with coal and stone traffic in connection with our Steam Motor Wagon Service'. It remained in railway service until 1921 when it was sold and replaced by another 4CD, No.33990. No.26479 was far from finished, however, and continued in private hands until it was last licensed in 1946.

As to No.26063, this remained in Garretts' hands for some weeks after its return from Smithfield and again seems to have been sent out 'on approval', in this case to the well-known showman, Jacob Studt, then at Victoria Hall, Southampton. Conclusive dates are once again lacking but spares for it were entered into the order book on March 20, 1907. As delivered to Studt the engine had no dynamo bracket but it was fitted with a 15" brass steering wheel; 12" brass brake wheel; brass wheel caps and a full length awning, the rods of which were brass sheathed. Though the order book refers to the steering and brake wheels as 'brass' references on Works drawings make it clear that they were brass plated. It was not until 1920, long after Studt had sold it, that an extended smokebox and dynamo bracket were fitted. Before he would actually purchase it he insisted on certain changes being made to it, the most important of which was the provision of machine cut gears in the final drive. Also requested was an improved interlocking arrangement on the gear change mechanism, the removal of the exhaust steam superheater, amendment of the 'starting gear', and the one pint Manzell oil pump being replaced by one that was 'silver plated' (or more probably nickel plated). All of this work was done without charge. Studt did, however, pay for the fitting of a dual water lifter enabling water to be lifted into a water dandy as well as into the engine's tanks. The tractor was finally invoiced to Studt on May 25, 1907. Even as amended it seems not to have pleased its purchaser as by March, 1909, it was traded in to Clayton & Shuttleworth, who resold it to William Freeman of Southampton. Two years later it was again sold, this time to Abraham Coneley of Ringwood, Hants, in whose ownership it was named *Invincible.* The tractor had a long life with the Coneleys, and was last licensed in 1942. On such evidence one must conclude that despite its not having come up to the expectations of Jacob Studt it was a sound and useful engine. A second tractor had been ordered by Jacob Studt at about the time the first was delivered but this was cancelled.

Whilst no drawing is known to exist of what exactly was done to the starting gear on Jacob Studt's tractor, inferences can be drawn from changes shown on drawing CR9519, prepared for a subsequent tractor No.27185 built for John Studt who had asked that the 'starting' (i.e. throttle and simpling valve) arrangement should be amended on the same lines as those used on No.26063. As the ports were originally designed, on drawing CR8343 (Fig.29), it was possible to run the engine on high pressure steam to both cylinders. It is quite likely that Studt's disapproval stemmed from the ease with which a heedless driver could thus abuse the engine by running it double high for a long period. To avert the possibility of such reprehensible behaviour, the port for the exhaust to atmosphere was blocked with a bar of soft copper hammered in, in conjunction with a reduced travel of the valve. Instead of the lever, which worked in a vertical quadrant, being fully forward for 'start' as in the general design the travel was restricted by an amended quadrant to the 'shut' and 'compound' positions only. To allow high pressure steam to be admitted to the low pressure cylinder, solely for starting, an external starting valve (or, as the drawing referred to it, a 'pop' valve) was mounted on the cylinder block. This is visible in Fig.27, and appears to be a plug cock rather than the later spring loaded type.

The use of shrouded gears for tractors generally was soon to be superseded by the fitting of gears machine cut from cast steel blanks. Whilst it might be rash, on the surviving evidence, to say which was the first tractor after Studt's to be equipped with machine cut gears it can be said with reasonable certainty that the change took place in 1907, and came into effect as soon as the stocks of shrouded gears were used up.

If the fact that the North Eastern Railway chose to replace its pioneer Garrett tractor when life expired with another of the same make suggests a thoroughly satisfied client, such marque loyalty was even more in evidence with No.26518, the next to be sold. On September 6, 1907, this went to John Harkness & Co. of Belfast. Harkness had been in business as a cartage contractor using horses since 1870 or thereabouts, but in 1907 took the plunge into steam traction, buying his first Garrett probably as a result of knowing W.E. Bullick, the Belfast agent for Garretts. The engine had been built for stock and was completed in June, 1907, so that it was available for rapid delivery. Apart from the fact that the flywheel seems to have been cracked on the journey out, needing a free replacement on September 19, it appears to have run into no problems. A new brake spindle was sent in December, 1907, but otherwise the tractor settled straight into work becoming the first of fourteen Garretts to be operated by Harkness. Indeed, apart from Wallis & Steevens tractor No.2926, which was bought second-hand, and briefly a Fowler 8NHP road loco, the firm ran only Garretts until the advent of diesels in the late forties and fifties. Even then the steam tractors and the road locomotive continued alongside them for another decade.

George Thurlow & Sons of Stowmarket, Suffolk, whose role in selling Garrett traction engines, rollers, and tractors in the home market was referred to in *Garrett Traction & Ploughing Engines,* next sold No.26712, another tractor made, it would appear, for stock, to Lewis Brown, the owner of a brick-field at Headley, near Newbury, on November 11, 1907. A further tractor from stock (No.26722), also finished in November, 1907, was bought by James Wood & Son of Boughton Monchelsea, near Maidstone, Kent. James Wood was one of the breed of high class country builders, now virtually extinct, capable of building for the owners of large landed estates, and he was an early user of steam transport, having earlier been the owner of a Coulthard steam wagon. The sale of this tractor, in February, 1908, was arranged by E.W. Rudd, the London haulage contractor who played a part in the design of the first of the Garrett overtype steam wagons. Rudd had introduced steam wagons into the business of hauling frozen meat from the London Docks to Smithfield Market, at first using Coulthards, for which he secured the London agency, but he lost interest in them after their take-over by Leyland. He had his first Garrett overtype wagon in July, 1909, and then went on to acquire a considerable fleet over the years, but his very first Garrett vehicle was a 4CD tractor, No.27088, delivered to him as a demonstrator on the last day of May, 1908, and equipped with wood-block rear wheels. These wheels (to drawing CR10011) were broadly similar to those fitted to No.3 tractor 25921 and were blocked with end-grain elm giving a diameter overall when new of 5' 1½". Although Rudd managed to arrange a number of eye-catching demonstrations with it, including a spell on hire to the Army for gun haulage, (for which purpose the wood-block wheels were temporarily exchanged for a standard set), the tractor did not fit in with his own business arrangements as well as wagons, and in February, 1911, he sold it to Crawley Bros. of Hadstock, near Cambridge. Later, curiously enough, firms like Union Cartage and Cornells, engaged in the Docks to Smithfield Market meat run in the 1930s, reverted to the use of tractors and trailers, converting units of their steam wagon fleets into weight-box tractors for the purpose, although this was more a consequence of the vagaries of taxation than of operational advantage.

Wood-blocked wheels had already been fitted to a slightly earlier tractor, No.27050, which was with the Lowcock Commercial Motor Co. of Gaythorn, Manchester, between February, 1908, and the following December when it was taken back into Leiston stock. Lowcock probably had this tractor merely for demonstration purposes as a note in the order book under the date January 20, 1908, says 'To remain the property of R.G. & S. Ltd. until sold.' Subsequently it was disposed of to the French agent Th. Pilter for his Algiers office. Before dispatch on July 5, 1909, it was modified to comply with French boiler regulations and amended to cater for direct ploughing. The changes involved the substitution of steel shod wheels 1' 8" wide at the rear, and 9" wide, with detachable angle rings, at the front, and the fitting of a special plough draw-iron on the tender.

The second showman's tractor to be delivered (No.27185) went to John Studt at Cardiff in August, 1908, equipped with a dynamo bracket and rails. As noted earlier, Studt asked for modifications to the steering drum and for changes in the starting gear. Other than this it was standard in the mechanical parts though fitted with a full length awning on twisted brass columns and embellished with stars on the motion plates, brass (or brass plated) wheels to steering and brake, and copper pipes throughout. Studt had the tractor painted in crimson lake with lemon chrome wheels. On the nearside of the roof he had the fascia lettered 'JOHN STUDT FROM THE CITY OF CARDIFF' and on the other 'JOHN STUDT, PULLMAN DINING CAR MOTOR'. For whatever reason he was not wholly pleased with it and ordered instead an engine which was similar but fitted with a Colonial firebox. This was sent to him as No.27884 on October 28, 1909. No.27185 was accordingly taken back by Garretts and resold to Mrs. E. Smith to whom it was consigned at the Ferry Hotel, New Passage, Devonport, with the name *Teddy Bear*. By about 1924 it was owned by William Thurston in Cambridge, but in the 1930s passed out of fairground use, seeing service with at least four subsequent owners, before finally being scrapped in 1942.

Studt's second engine was even more elaborately turned out than the first, having the reversing, starting and change speed levers all plated. The simpling valve was altered as on No.27185. Other changes specific to this engine were a cock in the equalising pipe between front and back tanks; the fitting of a sheet iron coal rack on the tender; a bell-mouthed drawbar; a fixed front draw-lug; and the placing of two intermediate awning supports in front of the flywheel rather than behind it, probably because of the greater length of awning that resulted from the use of the Colonial firebox. By 1914 it was owned by George Witherstone of Hereford, and in 1919 moved to White Brothers in Cardiff who called it *St. Tydvil.*

The strength of the Garrett overseas connections that had been very noticeable with the sales of traction engines was again demonstrated with the 4CD tractors, perhaps even more remarkably since the tractor was a machine designed specifically to comply with a British law not reflected in overseas legislation. Some of these sales necessitated extensive design modifications, something which a manufacturer of 1908 was much less reluctant to undertake than his present day equivalent. No.26969 for Grimaldi's Milan office was ordered with rear wheels 5' 0" diameter x 1' 2" wide as opposed to the standard 4' 6" x 10", and front wheels 3' 7" instead of 3' 4". The driving axle diameter was increased to 4⅛". As noted earlier this modification became standard later in 1908. Grimaldi went on to sell Nos.27112, 27170, and 27620, the latter with right hand steering but otherwise these were all fairly standard. Perhaps more remarkable was the sale by Mattos of Buenos Aires of sixteen examples, all mounted on Colonial boilers, to the Ministerio Obras Publicas, in the same city, beginning with Nos. 27367 - 27381 inclusive in February to April, 1909, an order which had temporarily eclipsed the building of standard tractors for stock.

However, building for stock was soon resumed and it was a tractor from stock, No.27342, that was modified to fulfil the order for a showman's tractor from Charles Hewson who travelled a bioscope show in industrial Scotland. Whilst Hewson was obviously acquainted with the tractor supplied to Studt, the details he requested for the full length awning on six twisted brass columns differed from Studt's in having the side boards standing above the eaves as a gutter each side. The dynamo bracket and rails were as supplied to Studt, and Hewson particularly required that the starting valve should be arranged as on Studt's engine. He further ordered a Manzell oil pump, twin Baines water gauges, a Willcox double pattern water lifter, and a 10' 0" extension chimney. The decoration was elaborate embracing not only stars and rings on the motion covers together with the brass name plate *Pride of Scotland* but also intricate lining out and the lettering *Hewson's Animated Pictures* on one side of the awning and *Hewson's Electric Biograph* on the other. It has had a long life, having been converted to a roller and been used by Donald C. Stewart of Bridge of Don, Aberdeen, from 1933 until the 1970s when, having become a convert to steam preservation, he put it back into nearly its original form. After he sold up it had a number of owners before going to the United States from whence it migrated to Australia where it is now handsomely restored in the ownership of Craig Robson of Sydney.

Meanwhile the way that No.26518 had stood up to hard work in the hands of John Harkness in Belfast had excited considerable interest in the area. No.27356 had been sold to William Lennox, a round timber merchant with an office in University Street, Belfast. Like many other timber hauliers later, Lennox was pleased by the generous size of the Garrett tractor's winding drum capable of holding 50 yards of 2" circumference (¾" diameter) wire rope where many of its competitors could only take that length of ½" diameter rope. Few tractors went without some modification to suit the customer. His, for instance, had a strengthened front axle. This tractor, in its turn, sold another. Robert Corry, the prominent Belfast building contractor and neighbour of Lennox in University Street, had No.27814, also in July, 1909, followed by another (No.27937) in October of the same year. At Tanderagee, in County Armagh, Thomas Collen had No.28219 in December, 1909, and in August, 1910, Samuel Rea of the Steam Saw Mills, Antrim, had No.28391. This went with 75 yards of ⅝" diameter wire rope on the drum. Obviously Rea went in for long pulls! Meanwhile Harkness himself had bought two further tractors (Nos.27655 and 27716) in May and July, 1909. Harkness was much involved in moving material for use in the shipyard of Harland & Wolff from Belfast Docks into the yard itself. Harland & Wolff were responsible, inter alia, for building all the vessels operated by the Oceanic Steam Navigation Company Ltd. who traded as The White Star Line. At the time of which we are speaking the chairman of Harland & Wolff was Mr. William J. Pirrie (afterwards Lord Pirrie) who was also a director of White Star. The White Star liners, *Celtic, Cedric, Baltic,* and *Adriatic* were built at Belfast between 1900 and 1906, followed by the *Laurentic* and *Megantic* and then by the *Olympic* and *Titanic.* It was the traffic created by this volume of work that was the main foundation of the Harkness business, and the difficulties of conducting it were outlined in a letter written by John Harkness to Garretts in January, 1912:

> The developments at the Belfast Harbour have caused the iron traffic for the ship-builders to be discharged on to newly re-claimed property, the approach to which is a very marshy nature, whereby our horses were unable to haul the traffic to the ship building yard. Your tractors have proved an invaluable asset in this connection, using them as we are doing in the tracing of the horses on these soft roads, and but for them the traffic would not have been removed. Again we are using them for hauling material for the large White Star liners, some of which measure 95 feet long, this also horses are incapable of doing by reason of the long trail on the ground. We are also using them hauling cables up a shute on to the lorry, by a wire rope on the drum. This work was formerly performed by five trace horses, and we found this very injurious to them. Your road locomotive also proves very valuable in our business. Pieces of machinery weighing up to thirty to forty tons can be safely and satisfactorily hauled up by us without any possibility of delay and accident. By having your tractors and road locomotive we are placed in an enviable position as compared with our competitors, because we can undertake the delivery of the machinery to a considerable distance outside Belfast and to many places where the Railway Companies are unable to compete. A decided advantage is gained by being enabled to remove machinery from the spot where discharged to the actual site where required to be delivered.

During the same period No.27352 (December 11, 1908) was sold to Coupe Bros, the heavy hauliers in Carlisle Street, Sheffield. Whether news had travelled from Belfast that the 4CD was a useful tool for heavy work or whether Teddy Ballam's sales organisation had interested them from scratch we are now unlikely to know. The only modification they requested was to the rear wheels i.e. to be 12" wide and shod with wrought iron, as opposed to steel, strakes. This tractor lasted in their hands for 22 years in the heaviest and roughest kind of work. Six years later they bought No.32631, which survived until the firm was absorbed by Hays Wharf Cartage Co. Ltd., a railway owned company, when the railway companies made their move into heavy road haulage in the 1930s.

Prior to 1914, when many County Councils were moving away from the use of contractors for road maintenance to the creation of direct labour forces of their own equipped with steam tractors and rollers, Aveling & Porter secured extensive sales in this market. Garretts, by contrast, made only one such sale, that of No.27717 to the County of Lanark in June, 1909. Why this should have been so is not readily explicable as there was little to choose between the tractors as to specification and workmanship. One concludes that the years of practice accumulated by Aveling & Porter in selling steam rollers to local authorities had made them very well acquainted with the ways in which local councils and their officers could be steered toward their products.

Customers in Garretts' home territory of East Anglia were also relatively slow to take up the new tractor, each potential buyer perhaps holding back to see how someone else had fared with one. The deadlock was broken when No.27886, built for exhibition at the 1908 Yorkshire Show, arrived back at Leiston in its 'local show' livery and polished copper pipes. Noah R. Etheridge of Blythburgh Lodge, Blythburgh, a few miles up the road to Lowestoft from Leiston was inveigled into the Works to look at it. A flamboyant character, given to wearing Stetson hats, Etheridge had a threshing and haulage business as well as his farming activities and was well known at Garretts. The tractor interested him but some haggling over the price took place. Eventually, sworn to secrecy, he was sold it at a 'special confidential cost price' of £375. In return he allowed it to be used for publicity purposes and a photograph of it at work threshing appeared in a number of the firm's catalogues. In the course of a letter written to the firm early in 1912 he wrote in appreciation of this 'after 25 years constant business transactions with your firm I very deeply appreciate the kind and gentlemanly way you have always treated me'.

Out of respect for the greater speed of the tractor he had three 6 ton trailer wagons on springs constructed by Garretts to his own specification which included artillery wheels with brass sleeves in the hubs in place of the cast iron boxes that did duty in most traction wagons. In 1912, mentioning that after 2½ years of hard wear they were practically as good as new, apart from damaged paintwork, he commented, obviously with feeling, 'stone hauling is very rough work!'

It seems that in service the first batch of tractors revealed two weak spots. The first of these was in the gearing. The North Eastern Railway tractor, No.26479, is noted in the engine memorandum book as having had new gearing, though nowhere have I been able to discover where the damage actually occurred. The gear train was redrawn on drawings CR10247 to 10252, entered in the drawing register on May 6, 1909, but never issued as working drawings. The drawing MR9014, for the original gearing, was formally replaced by MR11456, from the board of Tom Jones, issued to the Works on August 8, 1910, but probably recording what had been going on for some time rather than initiating a change as from that date. These changes increased the pitch of the spur ring and its mating intermediate shaft pinion from 1½" to 1¾", altering the ratio from 18:86 to 14:64, thus increasing the speed by about 5% and necessitating modified centres between the shafts. In Etheridge's case the use of the tractor for rough work had made this shortcoming in the gear painfully apparent but he was delighted with the new heavier gears which he characterised as being 'as stout and strong now as 6HP tractions of many makes'. He went on to add:

> After 2½ years wear, the tubes, firebox &c., seem to be as good as new. This engine has hauled thousands of tons of road material each year, 10 to 12 tons each journey, over all our roads. The engine drives a 4-ft. 6-in. new threshing machine of your make, and elevator, in fact plays with it, threshing 100 cos. of barley with 3½ cwts. coal, and 206 cos. oats in 10 hours, with 4 cwts. best Portland hard weighed in cwts., and in each case got up steam as well and had some to spare. My son has also taken 100 cos. of oats to Lowestoft, 15½ miles each way, and brought back 5 tons herring guts on 4cwts. coal for the day (31 miles in all). He has also carried 60 tons granite two miles in 6 journeys on 4 cwts. in one day, and had some to spare (24 miles).

One hopes that the wagons were well washed out between the herring guts and the next load of oats! A comb or coomb (the plural of which was commonly abbreviated to 'cos.') was a volume measurement for grain equalling two bushels. A quarter was originally a volume measurement equal to eight bushels. About the turn of the century a quarter of wheat was standardised at 516 lbs.

The other weakness appears to have been in the carriage jaw. Harkness, for instance, had a whole new front axle assembly for No.26518 in October, 1908. At the same time he had new rear wheels, still 4' 6" in diameter but 12" wide, with wood strips between the strakes, presumably to reduce noise and vibration. For all these items combined he paid £50.8.0. The result must have been satisfactory as not only he but his neighbours as well went on to invest in further 4CD tractors. The North Eastern Railway tractor (No.26479) also had a new forecarriage as did No.27088 owned by E.W. Rudd. Each of these clients paid a nominal £3 for the new assembly, and there may have been others not traced. Exactly what the alteration was has also proved not to be traceable, and was probably effected by a drawing office sketch and a small modification to the foundry patterns.

About this time, c.1910, Garretts began to canvass the idea of using the 4CD tractor in this country for general mechanised farming in much the way that later generations of farmers were to use the internal combustion tractor and, probably, as the Uganda Company had intended to employ No.25646. The first tractor known to have been sold for use in this way was No.28370 bought in February, 1910, by T.J. Mugleston who farmed Hammonds Farm at Stapleford Abbotts near Romford. With it he had one trailer wagon, capable of carrying about 10 tons, used to cart wheat, varying from 30 to 35 quarters at a time according to the state of the roads, to the station, 4½ miles distant and involving 'two rather steep hills', at least by East Anglian standards. Return loads included coal and manure. The Autumns of 1910 and 1911 being very dry, he used the tractor for direct haulage of a four-furrow Ransomes plough, reduced to three in the heaviest soil. He also did chaff cutting, and a certain amount of grinding and sawing. The tractor was ordered with 4" wide rear wheel extension pieces and angle rings but otherwise was standard. The Muglestons used it thus for some fifteen years so reasonable success must be presumed, but such examples remained rare.

In the same way the idea of using a tractor for hauling and powering a threshing set caught on to some extent but never generally. Since the threshing drum required the same amount of power to drive it irrespective of the power source it followed that when a tractor was used it had to run somewhat faster than say a 6NHP traction engine. Consequently the exhaust was fiercer and the fire brighter, tending to the formation of clinker and pulling into holes if not kept fired in small quantities at regular intervals. The water, too, was in and out of the glass rapidly and needed constant watching. Nevertheless, with a driver of the right temperament, the tractor could thresh as well as most 6NHP tractions and probably on a little less fuel, though because it was so continually in need of the driver's attention the latter had less time to spare for other jobs such as looking after the sacks, which drivers of bigger engines very frequently did. On the other hand the tractor could be coaxed into awkward stack-yards, along narrow farm lanes, and over suspect bridges where a larger engine could not have been taken. Within perhaps fourteen months of Etheridge taking his engine, seven tractors had been sold to firms who included threshing in their activities - the tractors concerned were Nos.27948, 27960, 28220, 28258, 28370, 28397, and 28745. In the case of No.27960 for J. Lucas of Holsworthy, Devon, it was provided with a belt roller on the front tank to enable the belt to drive a pulley 'only 4' from the ground'. It is suggested that these tractors were probably not employed wholly on threshing but rather that they doubled up at threshing along with road haulage.

Despite amendments made to individual tractors to suit the requirements of customers the basic design had remained little changed apart from the revisions to the gearing, but in early 1910 superheating and the use of piston valves were introduced as an option to purchasers. Much detail of the firm's interest in the use of superheated steam, both in portable engines and in steam wagons, as well as, to a limited extent, in traction engines, has already been given in my earlier books. Nevertheless one or two remarks may be added by way of elucidation. There are two main reasons for the consideration of superheating in a steam engine. One is to increase the effectiveness of the engine. The saturated steam that fills the steam space of a boiler loses pressure as it loses heat or vice versa. Thus where an appreciable length of steam pipe or steam passage occurs between the regulator valve and the steam chest the loss of heat that occurs to the walls of such conduits means that the pressure of the steam in the cylinder is less and also, since the ability of a given weight of steam to remain as a gas decreases as it loses heat and pressure, it will be wetter. With superheating, however, the steam which emerges from the boiler as a gas has further heat imparted to it over and above that necessary to convert it from water to a gas and hence can suffer heat loss without loss of pressure or increased wetness. This factor could be, and was, used to increase the effectiveness of a railway locomotive but in our more limited field of interest it was vital to the effective working, for instance, of an undertype steam wagon in which the length of steam pipe resulted in a serious diminution of power output by reason of heat loss and the resultant pressure drop when wet steam was used. This is one aspect of superheating. The second aspect is the ability to increase the overall thermal efficiency of the engine and by this means to achieve economy of fuel. In essence this can be achieved by increasing the temperature of the saturated steam by passing it through a superheater utilising heat that might otherwise go to waste with the flue gases. Both or either of these factors might have been in the mind of a designer or manufacturer contemplating the use of a superheater. In the case of Garretts c.1909/1910, being principally concerned with overtype engines where the steam passages were the shortest possible, the quest was mainly for thermal efficiency expressed as a saving in fuel costs.

Schmach has left no records of how his train of thought on superheater design progressed. That he kept abreast of the latest developments in railway and industrial steam practice there can be little doubt. One conjectures that the successful Schmidt system of superheating whereby a group of large bore firetubes contained long superheater elements extending from a header in the smokebox nearly to the firebox was considered and rejected, not least, one surmises, because the boiler of a traction engine or tractor could have been considered too small to accommodate such apparatus, though it was later to be applied successfully by Continental makers to ploughing engines. The choice of the Leiston drawing office fell upon the use of a tubular coil superheater housed in an enlarged base to the chimney. The smokebox superheaters applied to portable or semi-portable engines working for hours at a stretch on steady work could be shown by test to bring about economy of fuel and water, from which the inference was drawn, not really justifiably, that a similar economy could be achieved in the much more variable conditions of work of a steam tractor.

Two superheated tractors, Nos.28068 and 28076, the former on a Colonial boiler to cope with inferior fuel, were put in hand more or less simultaneously. Both incorporated the coil placed on an enlarged circular base to the chimney (see Fig.36) and carried a redesigned cylinder block incorporating piston valves, the latter used in all subsequent superheated tractors. Each was an exhibition engine turned out in show finish. The Colonial boilered example was destined for Mattos, the agent in Buenos Aires. Direct ploughing was clearly in the designer's sights as it had a double drawbar, the lower level for ploughing and the upper for road haulage, and 5' 0" diameter x 14" wide rear wheels equipped with malleable iron strakes 1¾" high of inverted 'V' shape, resulting in the outside diameter of the wheels increasing to 5' 2¼". The front wheels were increased to 3' 7¾" diameter by having a 2" x 1¼" cross-section additional tyre. Whether or not this tractor was intended for a specific show or for general exhibition and demonstration is not known, but at the end of its exhibition duties Mattos must have had a buyer ready for it as it never returned to Leiston. Even more encouragingly he ordered a second tractor (No.28768), generally similar and amended only in that it had an extra strong spur wheel and four bevel pinions in the differential. The case differed with No.28076. When finished it was shipped in early April, 1910, to E. Weber, the Belgian agent, to appear at the Brussels Exhibition. Apart from the superheater and cylinder block it did not differ from the standard design. Sad to relate it found no favour with Belgian buyers, and in consequence it returned unsold to Leiston in December, 1910. Fortunately for the firm it was there only long enough to have the necessary alterations made to suit a home customer found for it by Orlando Ralls, the company's agent based in Pershore, Worcestershire. The man whom he had succeeded in interesting in it was L.P. Stokes of Down Farm, West Kington, Chippenham, who asked for the same type of double drawbar as had been used on the Mattos tractor. He further wanted 5' 0" x 12" main wheels and 6" detachable steering angles to the front wheels. In addition he ordered a set of 20 'spudlets' for the rear wheels, and a spring seat for the driver. By quick work the tractor was ready for despatch by January 7, 1911. It is believed that the coil superheater was soon discarded but the tractor itself had a useful life, running in Stokes' hands for some nine years and eleven more for John Snow & Co. of Glastonbury, Somerset.

A further ploughing tractor, No.29783, to the same specification as No.28768, the second supplied to Mattos (but without the malleable iron strakes), was sold via James Macalister of Invercargill, New Zealand, at the end of September, 1911, having been shown in July at the Royal Show. Unlike Mattos, however, Macalister found no reason to repeat the order, and his next tractor, No.31126, was a No.4 single.

When the system of drawing lists (or specifications) was first set up at the beginning of 1908, two drawing lists for the single cylinder and compound tractors, numbered 9 and 10 respectively, were completed on March 26, 1908, the work of Thomas Jones. Although no major revisions of principles took place there had been sufficient changes for new specifications, Nos.34 and 35 - for the 4CD and No.4 singles respectively - to be issued on March 24 and 25, 1909, only to be replaced by Nos. 36 and 37 on April 19. The compound specification was not revised again until No.177 was issued on July 19, 1915. The sales of the tractors for direct traction ploughing were deemed of sufficient importance to merit the preparation of a drawing list of their own, No.48, completed on September 2, 1909, but unlike lists Nos.36 and 37 it was never superseded, though doubtless updated from time to time.

The more important changes that had taken place in the comparatively short period covered by this chapter can be summarised as follows:

[i] Changes to the cylinder casting (CR1), incorporated in a new pattern designated CR1A, which were probably the most important and far-reaching of the amendments, and involved a drastic simplification of the steam passages. The arrangement (Fig.29) as originally designed, whereby the engine could be arranged to run as a two cylinder simple (albeit of differing cylinder sizes), was abandoned, together with the hollow regulator valve casting that necessarily went with it. Instead a single port was used, admitting steam to the high pressure valve chest. The exhaust from the high pressure went to the low pressure valve chest and the low pressure exhausted to atmosphere as heretofore. Thus one steam passageway was abolished and two were combined, with great benefit to the cost of making the castings. Moreover the simpler passages allowed freer and more efficient movement of the steam through the transfer passage. Simpling reverted to the arrangement used on compound traction engines with a bronze valve mounted externally on the cylinder block and actuated by a push rod.

[ii] The abolition of the inconvenient side-opening damper to the ashpan in favour of a front damper which, in turn, involved altering the design of the front tank.

[iii] The redesign of the front tank so that its floor was only 1' 9" below the boiler centre line where before it had been 2' 8", necessitating lengthening the tank from 1' 9" to 2' 6". This change would have placed the steering bobbin, had it remained on the front of the tank, too close to the front axle and too high, so that , of necessity, it had to be moved to below the tank with the carrying brackets attached to the tank floor.

[iv] The increase in rear wheel diameter from 4' 6" to 5' 0" with the lowering of the front axle in the carriage jaw to suit, thus requiring a new pattern (MR82) for the carriage jaw itself.

[v] The abolition of the exhaust steam superheater in the smokebox with the exhaust taken instead direct into the smokestand (casting MR33). The earliest chimneys had been flanged direct to the smokebox top with a light cast iron flange fixed to the steel plate of the chimney.

[vi] The strengthening of the gearing.

The improvements effected to the 4CD tractor by the combination of these changes increased the appeal of a design that was already well liked and the enhanced sales that resulted are dealt with in Chapter 6. Before this, however, it is desirable to look at the new designs of rollers that evolved from linking the basis of them to tractors rather than to traction engines.

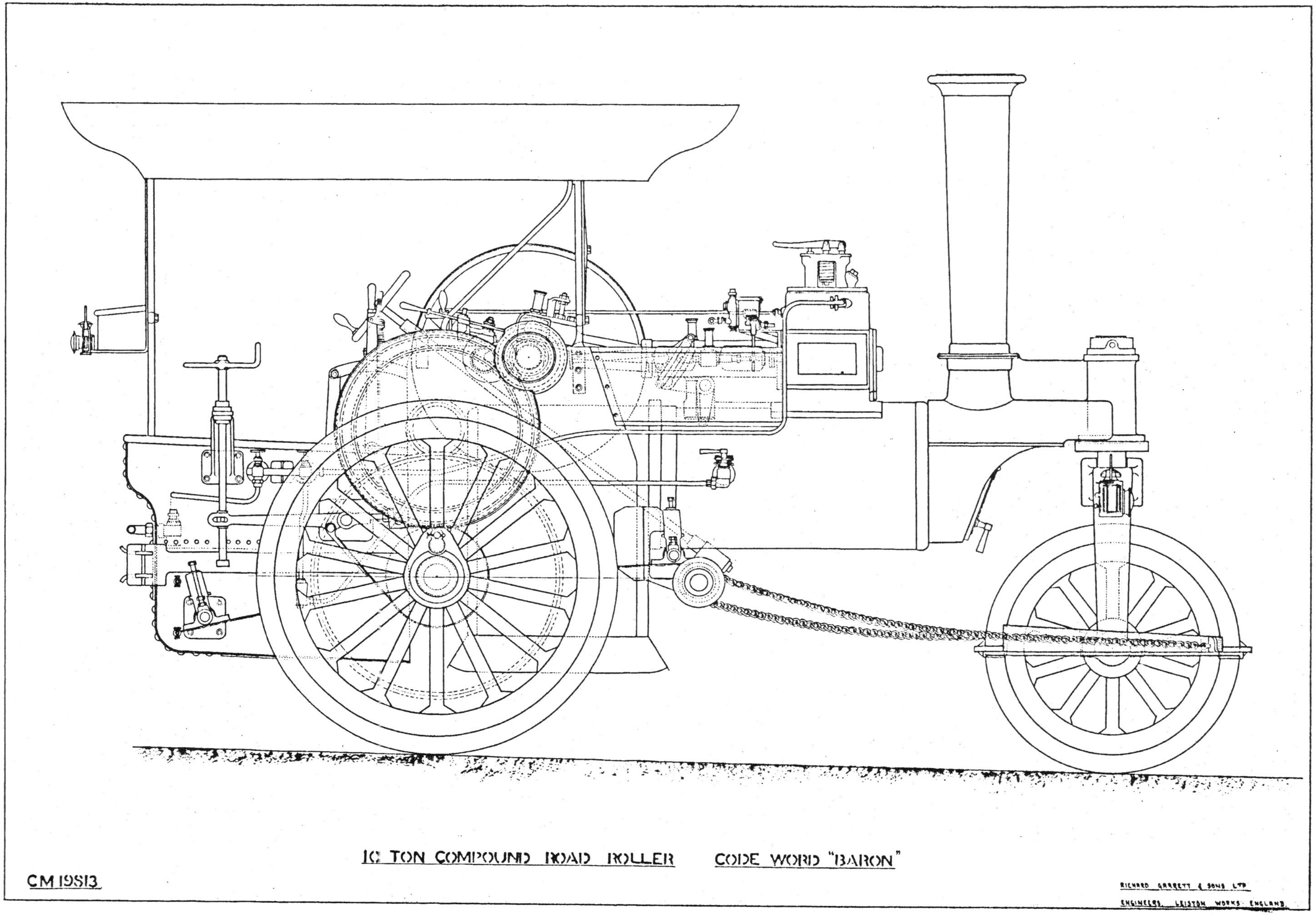

Fig.40. Drawing CM19813 showing the three-shaft design of roller code-named Baron.

Fig.41. The precursor of the Baron *series of 3-shaft compound 10 ton 'light' roller was No.27278 despatched in November, 1908, to the Italian agent Grimaldi.* [SRO]

Fig.42. The oldest Garrett roller in Great Britain, No.28408, (Aug. 3, 1910) new to Southend-on-Sea Corporation, and then requisitioned for War Service, was afterwards bought by Cumberland County Council. It is now owned by Richard Hullah. [R. Hullah]

Fig.43. No.29763 (July 20, 1911) was sold through H. von Arx of Zurich to Baudirektion des Kanton Bern, Switzerland. It was equipped with a coil superheater and piston valved block, and is now preserved in the Museum of Traffic, Luzern. [A.J. Martin collection]

Fig.44. *No.32648 (Feb.25, 1915), supplied to the Christchurch (NZ) Tramway Board, via James Macalister, was a special in many ways. It had a winding drum, differential, governor, Colonial firebox, springs to both rear axle and front forks, and all gears were non-standard . It is now owned by R. Miles of Rangiora, South Island.*

[R. Miles]

Fig.45. *10-tonner No.34706, new to Fairclough of Clacton in March, 1925, with the name* London Belle. *Subsequent owners were A.M.Emeny of Hadleigh, Suffolk, and his successor, Jack Gray. When the latter sold up it went to the Philp collection.*

[R.G.Pratt]

Fig.46. *Many of the Garrett three-shaft rollers latterly in service were conversions from tractors but No.34085 (Feb.8,1922) was one of a pair bought as rollers by East Suffolk County Council, becoming Fleet No.4. It is seen here between Darsham and Halesworth just before Christmas, 1945.*

[R.G.Pratt]

Fig.47. *The 5 ton three-shaft single cylinder roller based upon the No.4 single cylinder tractor, perhaps No.30059 or 30060 sold by H.F.Roll.* [SRO]

Fig.48. *Compound three-shaft convertible No.32049 supplied to Pilter's Paris office in December, 1913. New gears were sent for it in September, 1930.* [SRO]

Fig.49. *No.28069, the superheated version of the Baron roller exported to Mattos in Buenos Aires in March, 1910.* [SRO]

Chapter 5

The Three-Shaft Rollers

Although, as related in the previous chapter, No.25646, the first No.4 single, was sent out with the necessary parts to convert it to a roller, no one at Leiston seems to have envisaged any commercial possibilities in using the tractor design as the basis of a new range of rollers. The necessary stimulus in this direction probably came from an enquiry in 1908 by Grimaldi, the Italian agent, for a compound roller of about 10 tons *actual* weight. Since the correspondence relating to the matter was all discarded before I first had access to the Garrett records there is no knowing what Grimaldi's actual stipulations were, but what the drawing office put together for him, and what he actually ordered on July 3, 1908, was described in the order book as a 'special roller' constructed by using the boiler, engine work and gearing, including the differential, of a 4CD tractor, and adding to it a revised smokebox, newly designed cast steel swivel forks, and rolls with cast steel rims. These were all shown on drawings CM9667 and 9668, dated July 10, 1908. The completed roller, No.27278, left Leiston on November 19, 1908, at a cost of £426.

The matter might well have rested there had it not been for a visit paid in 1908 by Carl Remy to Thomas Cook & Son (Egypt) Ltd. in Cairo. Cooks placed a considerable volume of business in portable engines and threshing machines with Garretts, and it was appropriate that Remy, who had had charge of Garretts' overseas sales for many years and who had been appointed to the Board in 1907, should call upon them, timing his visit to coincide with the Cairo Exhibition at which they had a pavilion, displaying Garrett machinery. They put to Remy an enquiry for a compound roller of about 10 tons actual weight, economical to operate and compact in design. Doubtless he recalled the roller being produced for Grimaldi, and on his return to Leiston very likely brought the matter into the discussions he would have had with Schmach in his capacity as chief designer. Remy and Schmach were both German-speakers and because of this and of their long association as colleagues, extending back more than a quarter of a century, their discussions were probably franker and deeper than might have been the case in other firms when the sales director met the chief draughtsman. As a result Remy seems to have come to the conclusion that a new range of three-shaft rollers, built around the tractor design, had commercial possibilities and to have persuaded the board of directors that the time had come when it should be added to the products that Garretts offered their customers. By the time Cook's formal order was entered on December 28, 1908, not only was the new type of compound 10-tonner on general offer but sales literature for it had been drafted and a new telegraphic code-name *Baron* had been coined for it. Victor Garrett related the circumstances to me as an illustration of Remy's sound commercial judgement and understanding of the need for standardisation, but the three-shaft rollers never achieved quite the success that they seemed to merit.

Once established the venture proved to be sound, though slow to take off. From the manufacturing view-point, of course, it made sense since large parts of the tractor design could be moved *en bloc* into the roller. Thus the boiler, cylinder, engine work, crankshaft, intermediate shaft, and all gears, with the exception of the differential gearing, were as used in the tractors. As the rollers were only rarely supplied with either a winding drum or springing, this resulted in a certain amount of new design work and patterns. The designs for the rolls, the forecarriage head and front forks which had already been made for the Grimaldi order differed from the type used in the older 'heavy' rollers, that had been based largely upon the Aveling design. Steering movement in the new 'light' roller was catered for by a vertical pivot pin in an extended collar whilst movement in the other plane was catered for by a horizontal pivot pin carried in a jaw at the foot of the vertical pin and passing through the forks. This is illustrated by Fig.40. Using the tractor gearing entailed retaining the low speed of 1.67 m.p.h., considerably faster than the heavy rollers in which the slow speed was about 1.17 m.p.h. Though a difference of 0.5 m.p.h. may seem trifling, considered in relation to a speed of 1.17 m.p.h. it represents an increase of 43% so that the minimum speed in low gear, as well as the speed at the rate of the notional governed revolutions of the engine, was that much higher. Thus the new rollers tended to be a little fast for good rolling. The drawings were prepared either by or under the leadership of Thomas Jones, the draughtsman who had already been entrusted by Schmach with much of the work on the revivified traction engine design of 1895-96 and the associated 'heavy' (i.e. four-shaft) rollers as well as the 6 and 8NHP road locomotives and the 4CD tractors. Of these the 6NHP road locomotives were probably the least successful, in part because of the changes made to his original designs by his superior in his anxiety to keep down gross weights. No.27578, the first roller built to this design, was shipped on March 10, 1909, to Thomas Cook. It would have been nice to have been able to record that it opened the way to a torrent of sales of such rollers to Egypt but in reality this was not so, and when Thomas Cook ordered another roller, shipped fourteen months later, it was a 15 ton superheated single cylinder roller of a type that will be described later. It was not until 1914 that Cook had another three-shaft roller, No.32585, this time for the Assiout Municipality. The second *Baron* to be sold, No.27839, went on July 9, 1909, to H. von Arx & Co. of Zurich, and was noted as having cast steel rims to both front and rear rolls.

It was not long before a roller appeared based on the superheated version of the 4CD tractor, equipped with piston valves. The forecarriage head had to be redesigned to accommodate the superheater coil, resulting in what could arguably be described as the most uncomely design of three point roller ever produced. The first of these superheated rollers to leave the Works was No.28069 which went on March 9, 1910, to Mattos in Buenos Aires for the *Exposicion Centenario* in that city. Built as one of a pair it was soon followed in early April by No.28077, which was taken by the Belgian agent, E.Weber, to the Brussels Exhibition where it stood with tractor No.28076, sharing the latter's fate in that no purchaser came forward with the result that it was taken back into stock at Leiston at the end of the year. After a refurbishment and the provision of an awning over the tender it was sent out again on April 29, 1911, for Grimaldi to show at the Turin Exhibition. Grimaldi had better luck with it than Weber, and it never came back to Leiston.

No.29008, an 8-tonner supplied to Davos Town Council in May, 1911, through von Arx, also had the coil superheater and the piston valved block. This roller had cast steel rims and was equipped with a Bomford 2-tine scarifier for which a charge of £65 was made. Apart from the double spring balance safety valves needed to comply with Swiss boiler regulations and the fitting of double gauge glasses, enamelled at the back and with red lines, specified on all of von Arx's orders, the roller had other features which denoted a client determined to have the best. All the firebox stays were drilled, and all pipe work was in copper. Royal Show finish was specified and the roller bore the Davos coat of arms in blue and yellow. As with preceding orders for von Arx, this roller had a Mollerup oil pump, never popular at home where the Empire or the Manzel were preferred. Despite the fact that, to the purist, it may have appeared an oddity, this roller had a long working life and survives in preservation. Few other clients ordered their rollers with superheaters despite the firm's claim that the introduction of a superheater was likely to produce economies of the order of 25 to 30% in both coal and water consumption, a wildly exaggerated statement which would have brought retribution under the Trade Descriptions Act had it then existed. Most intending purchasers doubtless considered the extra cost of the superheater and the extra maintenance required outweighed any actual saving.

Meanwhile Ballam's sales staff had sold No.28408 (August 3, 1910) to Southend-on-Sea Corporation, who asked for special Royal Show finish. The roller was requisitioned by the War Department during the 1914-18 war and never returned to Southend, being sold by the War Stores Disposals Board to Cumberland County Council for whom it worked until 1954, when steam preservation was in its very early stages. Named *Elizabeth* it is now owned by Richard Hullah and is frequently seen at rallies, travelling long distances to and from them under its own power. The next two to be sold in numerical sequence were No.29007, which went to Muscate of Danzig minus front and rear rolls, in the established pattern for German sales, and No.29011 sold to the Municipality of St. Moritz via the agency of von Arx. The latter roller went in Royal Show finish set off by the coat of arms of St. Moritz and with a full length awning and copper pipes throughout. It was ordered with a damper in the chimney, common in Switzerland in preference to the usual plate. So far as this series of rollers was concerned, probably the best event of 1911 was the order by Jean S. Marcantonakis of La Canée, Crete, for a batch of five for use by the Cretan Government. In these rollers the ordinary firetubes were of brass, though the stay tubes necessarily remained steel. Marcantonakis' son was one of Frank Walker's 'gentlemen apprentices' at Leiston, becoming in the process a pronounced Anglophile.

As with so many Garrett products the rollers were far more popular abroad than at home. For instance, Burns in Howrah, beginning with No.30412 for Cawnpore Municipality in July, 1912, returned repeatedly with further orders. Thus in 1913 they were responsible for selling Nos.31557, 31651, and 31818, although when the Colombo Electric Tramways & Lighting Co. Ltd. ordered a roller they did so via Robert Brooks & Co. of 11 St. Benets Place, Gracechurch Street, in the City of London. Their roller,

No.32623 of November, 1914, on a Colonial boiler, had an Allen scarifier, and the woodwork of the full length canopy was specified to be 'treated with Solignum or equal to render it immune from white ant attack'. Another export to Greece, this time through the agency of Gregoriades Bros, was No.32665 (May, 1915) destined for the town council of Cavalla, although it was not built for the order but was already in stock at the beginning of the year. Thereafter no more of this series left the Works until 1920.

Besides the 10 ton *Baron* series, a parallel line of 8-tonners code-named *Prince* was introduced, commencing with No.30349 for the Ministerio Obras Publicas in Buenos Aires, via Mattos. This left on April 19, 1912. The reduction in nominal weight was brought about by reducing the thickness of the rims but otherwise the design followed closely on the *Baron.* This particular example, which had a short awning over the tender, was ordered with an Empire oil pump, Klinger gauge glass, and cast steel fork. The second of these nominal 8-tonners to be sold (No.30851 of August 8, 1912), via J.Chambers & Son Ltd. in New Zealand, was really a wolf in sheep's clothing in that although sold as an 8-tonner it was ordered with a winding drum and 50 yards of wire rope, rear rolls increased from 15" to 19⅞" on the tread, so as to bring up the rolled width from 6' 2¼ to 7' 0", and the thickness of the rolls increased from 2" to 2¼" at the front and 2" to 2¾" at the rear, i.e. although charged out as a *Prince* it was virtually the same as a *Baron.* This, too, had cast steel forks. An almost identical example (No.32491) was supplied in October, 1914, to Mount Albert Borough Council in New Zealand via Chambers. Returning to 1912 there followed No.31178 for Grimaldi's Turin office in December, 1912, and another example (No.31956) for Chambers a year later. This also differed a good deal from standard. Like its predecessor it had a winding drum and 50 yards of wire rope, but additionally it had a differential, was fitted with Pickering governors, and had a Colonial firebox. Chambers also ordered it to be provided with an Allen 2-tine scarifier. Orders, all for export, continued desultorily but were soon inhibited by the outbreak of war in August, 1914. Thus No.32224 was sold in April, 1914, to the Municipality of Cordoba through the London agency of Torromé, who acted also as the London confirming house for Mattos of Buenos Aires. No.32224 was followed by a further roller for the same client, although in complete contrast, namely the last 18-tonner to be made, No.32315.

The only example supplied in 1915, No.32648 sent on February 25, was ordered by J. Macalister in New Zealand for the Christchurch City Council Tramway Board. This was very much a special in that it had a Colonial firebox, all the gears were non-standard, and the hind axle was sprung as in the tractors. Spiral springs in the forks allowed limited vertical movement of the steering axle. Besides this it had a differential, a governor, and a winding drum. This roller still exists in New Zealand (see Fig.44) The year 1917 saw No.32922 shipped to Chambers in New Zealand at the end of October, with very much the same special features as No.31956. In the same month No.33021 was sent to the Royal State Railways of Siam at Bangkok. This was another roller built on a Colonial boiler, probably because it was to be fired on wood, and had a differential and a full length awning. Further, it was ordered with a detachable smokebox, although no parts to make it capable of being fitted out as a tractor were asked for. It was known to be still in use in 1930 when new rolls were supplied for it.

Apart from these rollers based on the compound tractors Garretts also offered 5/6 ton singles, code-named *Earl,* based on the No.4 single cylinder tractor. These were sold via Burn & Co. to Indian and Burmese customers, though whether or not they were developed specifically to meet Burn's requirements is not known. As drawn they were classified as 5-tonners and had cast steel rolls 1½" thick, but the first to be built, No.30992 despatched to Rangoon Municipality in November, 1912, had 2" thick cast iron rolls and was sold as a 6-tonner as was No.31148, three weeks later. Just over a year later, on the last day of 1913, No.31267 left as a 5-tonner, in the designed form and fitted with a cast steel fork. Nevertheless with No.31284, six weeks later, there was a reversion to the 6 ton form using cast iron rolls. However, when Burns ordered a '7/8 ton single cylinder roller' in 1914 what was sent was what the Garrett order book referred to as a 'Steel Earl' with an internal note 'our 5 ton'. Interpreting what Burn's orders needed was probably an interesting exercise as they fluctuated between three and four-shaft types for the same nominal weight. Because the competition for orders was so keen, as has been already noted, doubtless what was supplied was based as much upon what it might have been deemed possible to persuade the customer to accept as upon a consideration of nominal weights. The last *Earl* known to have been sold was No.33809 which went to the City of Bombay, through the agency of John Birch. Despite both rims and forks being of cast steel the nominal weight was described as '7½ tons'.

The three-shaft design, being in essence a tractor adapted as a roller, lent itself to convertible form. Indeed, in later years many tractors were converted to rollers. One of the first to be supplied new as a convertible was No.28372, ordered through Grimaldi for Ing. Arrigoni & Co., Milan, who also purchased the straightforward 12 ton four-shaft compound 28371. Both rollers left on April 9, 1910. In No.28372 the standard cast steel swivel fork was matched with a two part forecarriage head bolted together. The roller had a differential and a winding drum with 50 yards of wire rope. The tractor parts sent with it included the front axle and carriage jaw, steering chains, hind wheels 5' 0" in diameter and 12" wide, front wheels 3' 7¾" x 6", and a Pickering governor. A turnplate was permanently in position below the smokebox which was longer than standard. Arrigoni ordered two injectors and no pump. This convertible has survived and is preserved in working order at the Museum of Road Building, S.O.M.A., Ede, Holland.

Fig.50. *Although the superheated three-shaft rollers were far from handsome the palm for uncomeliness probably should be bestowed upon the four-shaft superheated type of which No.28049 was the first, exported on July 16, 1910, to Grimaldi's Turin office. It had been put in hand as a speculation and completed early in 1910 with Ramsbottom safety valves as shown. These were altered to spring balance before the roller went to Grimaldi, and it was then also fitted with an awning and Pickering governors.*

Fig.51. *10 ton single cylinder roller No.30857 (July 16, 1912) for the Public Works Department of the Swiss Canton of Neuchatel, sold through the agency of H. von Arx of Zurich. This was a four-shafter with 12" stroke, code-named* King. [Both SRO]

Fig.52. *No.29031 was delivered fourteen months earlier than No.30857 and still had the old type of fork and forecarriage head. Bought by Zachariah Fairclough (by rear wheel) of Clacton and named* Clacton Belle, *it was driven by Jack Faiers. It was unusual in having twisted brasses to the roof columns.*
[B.Essam collection]

Fig.53. *No.32076 (Jan. 1914) an 8 ton four-shaft single cylinder roller for the State Railway of Java.*
[SRO]

Fig.54. *A standard 10 ton single for export, possibly No.32137 (March, 1914) for Henninger of Darmstadt.* [SRO]

Fig.55. One of the four-shaft single cylinder rollers sold to Henninger of Darmstadt, delivered without rims to the rolls and minus scrapers and the like. Code-named Monarch *and described as a 5 ton it was simply a lightweight* Reading. *Probably it is No.28414 (August, 1910). It is shown here at Maldegem, Belgium.*
[D. Rayner]

Fig.56. The gear-side view of a 6 ton four-shaft roller, c.1909, the Works number of which cannot be stated with certainty, but probably is 27628.
[SRO]

Fig.57. The flywheel side of a similar roller, possibly No.32522 for Burn.
[SRO]

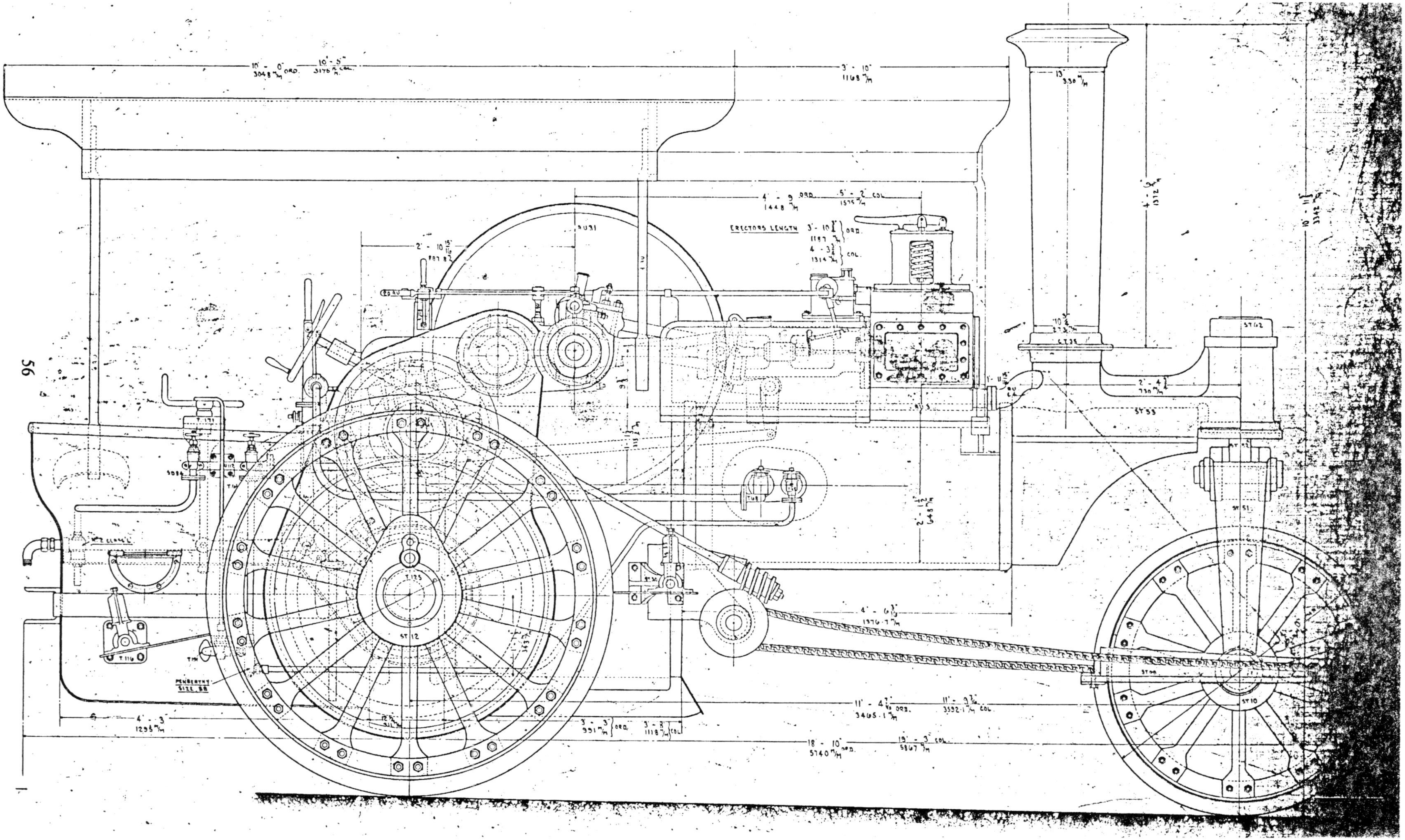

Fig.58. *Gear-side elevation (AU21657) of a 10 ton standard compound roller showing the swivel fork, as re-traced in 1924 for Ansaldo (see Chapter 9).*

Fig.62. *No.30987 (October, 1912) was bought new by Joseph Sanders & Son of Hanley Castle and named* Queen of Hanley Castle. *The load is an elm butt from the park at Endon Hall near Pershore.*

Fig.63. *A.H.Williams of Monmouth had tractor No.32632 in December, 1914, in the opening months of the first war. This shows it in new condition.*
[A.J.Martin collection]

Fig.64. *The war was still on (note the military figure on left) when this gear-side view was taken, but the tractor had acquired a scar or two, particularly at the front of the awning. The driver, too, seems to have aged somewhat. It is rare to have views of each side of a tractor, and so close together in time.*

Fig.65. *A.H.Williams, who was a hay merchant, also had 4CD No.30929 bought from Jones Bros. of Treharris, Glam. who had it new at the end of June, 1912. Here it is involved in actual hay-making. From the beam on the driver's face the copper vessel held by the mate must have contained the 'haysel' cider.*
[A.J.Martin collection]

Fig.66. *No.30085 of April, 1912, began its life with J.Brooks Taylor of Neath, Glam. hauling bricks, but by the time of this picture it belonged to Bristow Brothers of Hereford who were haulage contractors.*
A.J.Martin collection]

Fig.67. *Superheated 4CD No.32029 was one of the first with the revised design of superheater in the flat topped smokebox, and was bought new by H.Lancaster & Co. of Newent (Glos), one of the minority of saw-millers who hauled in their round timber with their own plant as opposed to having it done by contract. The splashguard over the crankbox is very noticeable.*
[A.J.Martin collection]

Fig.68. No.31209, (31 December, 1912) the superheated tractor upon which Noah Etheridge lavished such praise, passed to Taylor Bros. of Bildeston, Suffolk, about 1920, who owned it when this photo was taken.

Fig.69. During the 1914-18 war the shortage of horses and horse fodder led the Metropolitan Asylums Board to turn to steam tractors for transport to their fairly isolated hospitals. They had No.32985 in mid-January, 1917.

Fig.70. The original owner of No.32703 (May, 1915) was Alban Thomas of Caerleon, Mon. but by the time this picture was taken in the 1920s it belonged to Robert Wynn & Sons of Newport who had fitted it with new wheels (from Garretts) provided with solid rubbers. The cylinders forming the load were water filters for the Newport Corporation reservoir.

Fig.71. *Tractor No.32084 (March, 1914) was ordered by Thomas W.Ward Ltd. for use at their Silvertown depot. It was made for 200 p.s.i. instead of the standard 180 p.s.i. but otherwise was unremarkable in the engine work. However, it was fitted with Tangent wooden wheels, invented by the Tangent Wheel Works Ltd. of Clapham, London, but by then made by Bauly's of Bow. Whilst the fronts had mild steel tyres, the rears were shod with wood and rubber blocks, a serious attempt both to combat the noise and vibration on sett roads and to improve adhesion.*

Fig.72. *Also for a customer in the Thames-side Essex suburbs was No.32653 (Jan.1915) for E. & A.Shadrack, the East Ham coal merchants. They soon found steam wagons more convenient, and sold it in the following year to John Mowlem & Co. Ltd., the London building contractors.*

Fig.73. Even in 1956 when S.J.Rhodes photographed the John Harkness tractor No.33414 (Jan.1919) it was an anachronism in its city surroundings, notwithstanding having been equipped with solid rubbers and electric lights. A visitor encountering this outfit in a Belfast street must have wondered where he had wandered.
[A.J.Martin collection]

Fig.74. In June, 1919, Harkness had another 4CD *tractor, No.33545, seen here with driver David Ennis in 1959. The flat smokebox door was a local repair job but the new wheels with rubber tyres came from Garretts.*
[A.J.Martin collection]

Fig.75. The war had heightened the interest in mechanical farming when B.T.Kent of Shelford Bottom, Cambs bought No.32835 in 1916. Not only was it fitted with a governor but also a double drawbar attachment and steering angles to the front wheels plus 20 spudlets for the rear wheels to make it suitable for direct traction ploughing. Kent did not make a success of this so that in 1920 it went to Sydney Wilson of Orford, Suffolk, and in 1923 to Mrs.French at Birdbrook. It ended its life in George Taylor's scrapyard at Redbourn, Herts.

Fig.76. *A line-up of Ministry of Munitions tractors awaiting delivery to Kempton Park, Surrey, by road under their own steam. Note the longer chimneys fitted to many M-o-M tractors.*

Fig.77. *No.33225 was one of a batch of four Ministry of Munitions tractors delivered on March 12, 1918, but in June, 1946, when photographed in Haylings Road, Leiston, within 300 yards of where it was first erected, it belonged to W.Moore of Benhall.*
[R.G.Pratt]

Fig.78. *No.33128 was also a Ministry tractor delivered on March 12, 1918, and later belonged to R.F.Fenn of Frating, near Clacton. Unusually for a tractor not used on the fairground it had the awning rebuilt to full length.*
[R.G.Pratt]

Chapter 7

The Later Tractors 1912-1919

The history of the later tractors is virtually exclusively that of the 4CD since only one No.4 single cylinder Colonial tractor comes into the period of this chapter, and the No.3 tractors that would fall into it, datewise, have already been given mention in Chapter 3. The solitary No.4 single was No.31126, despatched from Leiston on November 19, 1912 for J. Macalister in New Zealand and was a special in a number of ways that suggest it was meant for winding and belt-work rather than haulage, i.e. no drawbar was fitted; a larger winding drum and 100 yards of ½" wire rope were provided; and there were four pinions in the differential instead of two. As no motion covers were required one presumes that New Zealand horses were deemed less neurotic than their British counterparts. Hind wheels 14" wide and fronts 6" wide were fitted. With this not very typical example the No.4 single cylinder bowed itself off the stage, only the twentieth example to have been built but the seventeenth to have been exported. Thus a tractor designed to satisfy the requirements of British law found most of its clientèle abroad.

The compound tractor, however, continued to gain acceptance. During 1912 sixteen examples were built, compared with the thirteen despatched throughout 1911, but the following year (1913) twenty-six were sold. A number of those who bought tractors in 1912 seem to have done so on an experimental basis. Thus Tom Everton, Junior, of Droitwich had No.30007 in March, 1912, it having been built for stock in the previous October. Apart from having a geared pump, the fitting of an *Empire* oil pump, and copper pipes the engine was of standard turn-out. By that time awnings had become virtually standard practice though officially they were to special order only. Everton had his tractor for some four years, doubtless comparing its money earning potential against what he could earn with his steadily augmented fleet of steam wagons and also, one would expect, comparing it with the Burrell tractor which he also owned. His choice came down mainly in favour of wagons, and in 1916, when tractors fetched good prices because of wartime shortages, he sold No.30007 to Peter Preece & Son of Gloucester. Later it was used on round timber haulage by C.H. James of Kingswinford and outlived the second world war.

The potential of using a tractor both for road haulage and on the belt became increasingly appreciated. Three tractors (Nos.30858, 30987, and 30988) were supplied with a roller on the front tank to keep the belt off the tank top. Of this trio Nos.30858 and 30988 were superheated, suggesting that their initial purchasers expected them to spend a substantial part of their working lives on the belt, where a superheater was most likely to improve economy of fuel and water. Whether or not they reached this conclusion independently or were influenced by Garretts' sales literature or the persuasion of the sales force one does not know, but the idea of superheating did gain a little ground at that time. When Noah Etheridge, pleased with the performance of No.30050, came to purchase a second tractor in 1912 he chose to have it superheated, though in other respects it was standard. No.31209 was delivered to him, on its own wheels, on the last day of 1912. In September, 1913, probably in response to a prompt from Teddy Ballam, he wrote of it, obviously when in a reflective mood:

> My son tells me that the Superheated Steam Garrett Tractor will haul quite 2 tons more than the Saturated Steam Garrett Tractor. He has been hauling this week every day from Framlingham Station to Badingham, near the White Horse Inn, about 4 miles, each tractor hauling 2 trailers, and doing 4 journeys per day. The stone came from Wymondham and was very heavy. The superheated tractor hauled 12 tons each journey, and the saturated could only manage 10 tons each journey as the roads are none too good. Of course he thinks the superheated steam type requires a better driver as there are more little things to remember and look after, but nothing which should worry any good traction engine driver. A careless driver, who cannot be depended upon, if it comes to that, would probably ruin a saturated steam slide valve type engine just as soon as he would one of your superheated steam piston valve type engines.

By 1920 it had been bought by Taylor Brothers of Bildeston, Suffolk, by whom it was used both for threshing and for round timber work. Afterwards, moving with the times, they used Garrett 3-ton wagon No.31225 converted to a weight-box tractor and Foden timber tractor No.13232 for the timber hauling. As a result the 4CD was not licensed after 1934.

There were, on the other hand, a number of tractors, like Everton's, supplied and used solely for road haulage. Tractor No.30085 bought by J. Brooks Taylor of Neath, Glamorgan, was used to cart the bricks from his brickyards whilst No.30959, sold new to John Griffiths of Bargoed, moved in 1916 to William Brown & Co. the timber and builders merchants of Ipswich for whom it worked on deliveries in the

Ipswich area for thirteen years or so. The tractor, No.31108, supplied to Tom Evans of Troedyrhiw, Glamorgan, in November, 1912, another haulage tractor, was one of those ordered with an oil splasher over the large ends of the connecting rods, suggesting that it was expected to be hustled along. Tractor No.31118 (February, 1913), supplied to the Devon Trading Co. Ltd. of Exeter, was a standard tractor, specially stipulated on the order to have no governors, and to be painted in traction engine dark green unlike the bulk of the tractors supplied about that time which were painted in chocolate colour. It lasted with Devon Trading until March, 1922, when Garretts accepted it in part payment for a wagon, selling it on at once to William Reynolds, the engine dealer in Bedford from whom it passed equally quickly to W. Price of Wasford, Herefordshire.

Although these tractors supplied for haulage had awnings over the manstand it was rare to find one with a full length roof, only two of the 1912 output being so equipped. One of these was No.31173 sent on December 12 to W.P. Williamson of Ahorey, Portadown, Co. Armagh. It was also distinctive in having a steersman's step complete with foot-guard to make room on the manstand for a second man, thus, one would have thought, throwing away the benefit of the legislative concession to 5 ton tractors, namely of allowing their manning by the driver alone. The other engine with a full length awning was probably the more interesting of the pair being the second showman's tractor to be supplied to a bioscope show. This was No.31025 *East Lynne* sold to Tommy Cottrell, whose permanent address was in Droitwich but who, at the time of purchase, was built-up on a semi-permanent basis in Leiston. No one in Leiston remembered exactly when he first came, most people putting the date in the years 1910 or 1911, and he was not there all the time. His bioscope show provided the first moving pictures ever seen in Leiston. Mostly he used a 'fit-up', sometimes on a meadow in Valley Road, but at other times on the land opposite the *White Horse* as well as occasionally using the hall of the Constitutional Club.

His purchase of the tractor probably marked the change from a gas illuminated projector to the much safer electrically lit type but no one at Leiston remembered this for certain. In much the same way as Hewson's slightly earlier tractor it had a dynamo bracket and rails mounted on an extended smokebox, had the stanchions of the roof columns brass sheathed, and was fitted with full length curtains to close it in all round when out of use. Other polished brasswork comprised axle caps, steering wheel, stars and rings on each motion guard, a ring on the flywheel, and star centres on the cylinder covers. It was also fitted with a 10 foot long extension chimney. Tommy Cottrell's link with Leiston was effectually severed when a number of local business men, including Frank Walker, Emerich Schmach, and J.B. Harrison from Leiston Works, financed and built a fixed cinema in the Aldeburgh Road, Leiston. Soon afterwards he gave up, and in 1915 the tractor was sold to Swales Forrest, then at Gravesend. The Forrests last used the tractor in 1932 and subsequently it was broken up.

Two 4CD tractors (Nos.30143 and 30968), both fitted with open tenders, were exported to Pilter's office in Marseille during 1912. The earlier of the two had a long life, and is now preserved in France. The following year he had two further tractors but with a difference. No.31511, which left Leiston on June 6, 1913, had the same characteristics as Cottrell's engine of the previous year but without the polished brasswork. Pilter asked, however, for a specially heavy flywheel. The ultimate client was Monsieur J. Broquet, owner of the *Grand Manège de la Vallée d'Eure*. M. Broquet had his tractor painted red. The second 1913 tractor for Pilter, No.31840, was finished in crimson lake with lining and decoration that was described in the order book as 'gilded'. The stanchions and cross stays of the full length awning were cased in twisted brass and the wheel caps were also of brass but there were no stars or rings. Again Pilter requested the heavy flywheel and, presumably to deaden the sound, a raw-hide gear on the pump.

Although the total of twenty-six tractors sold in 1913 represented a useful improvement upon the number sold the previous year, it was eclipsed by the thirty-six sold in 1914, the first of which, superheated tractor No.31934, left on January 31, although actually completed the previous November and shown at the 1913 Smithfield Show in December. It was purchased by Bell's United Asbestos Co. Ltd., and imported via the port of Lagos for service in Kudaru, Nigeria. The order specified that it should carry 100 yards of wire rope but otherwise it was standard. Sales of the superheated version reached a peak of eleven in 1914, just exceeding those of 1913 (ten), and overtopping the seven sold in 1915. In 1917 four were sold but other years failed to rise above two examples. The total score of superheated 4CDs was forty-six out of a total of 549 No.3 and No.4 tractors of all types.

The second to go was No.31979, finished during the last week of 1913 and sent by rail on January 31 to George Weller of Cranleigh, Surrey, who ordered it with the parts needed to convert it to a roller when required. In the event, it spent most of its early life on timber extraction which its purchaser obviously had in mind when he ordered it with 'large guide rollers'. Somewhat unusually it also had a full length awning and polished wheel caps. It worked for the Wellers until 1929, latterly in its roller form, until traded-in in part exchange for a 6 ton Advance roller (No.8028) with Wallis & Steevens, who sold it for scrap to Albert Maynard of Addington, Surrey, without its ever going to Basingstoke.

The superheating arrangements applied to tractors had undergone a revision as from No.31647. Although the cylinder with its piston valves was not changed the superheater coil in the chimney base was replaced by a design, similar to that used on the contemporary wagons, incorporating two cast steel headers and eight banks of 1 3/8" outside diameter tube, a total of 46.4 feet of tube, thus adding some 12 square feet to the heating surfaces, the whole arrangement housed in a flat-topped upward extension of the smokebox. On the working version of drawing BCR13639 (dated March 14, 1913) the coils were in two groups of four on either side of a central space to accommodate the exhaust pipe, but this is an amended version of an original layout in which two further coils occupied the central space, i.e. 10 coils in all which would have given 15 square feet of internal heating surface. Whether or not a boiler was actually built with these two extra coils is not recorded. A steam lance and flexible steam pipe was provided to help with the thorough cleaning of the superheater tubes. Keeping the superheater at all effective depended upon how well and how often the driver performed this exercise. No.31647, completed in July, 1913, to the order of James & Copeland, London agents for Powysland Plantations BEA Ltd. in what was then British East Africa (now Kenya and Uganda), was interesting in other ways also, not least in that it was stipulated that the ground clearance under the ashpan was to be 2' 3". This was achieved by increasing the rear wheel diameter to 6' 0" - with a width of 14" - and mounting the 4' 6" x 9" front wheels on an axle cranked down at each end. These front wheels were provided with detachable steering angles whilst 6" wide bolt-on extensions with angle ring fixings were sent for the rear wheels. The water storage capacity was increased from 145 to 201 gallons; all pipework was of copper; a special double water lifter was fitted and 26' 0" of additional suction hose was supplied.

Nos.31777, 31780, and 31800 supplied respectively to Hough & Co. Ltd. of Kings Heath, Birmingham, B.W. Hillen of Maldon, Essex, and A.& R.Brown of Colinsburgh, Fife, had superheaters of this type as also did No.31901, sold at the end of 1913 to the Express Motor Transport Co. Ltd. of Pershore, with which Orlando Ralls, the Garrett agent in the area, was connected. Bearing the name 'Express No.5' the tractor, like Weller's, had a full length awning, stronger rope guide rollers, and 75 yards of wire rope. Other special items ordered included a wheel valve in lieu of a plug cock on the steam supply to the water lifter, a guard over the pump gears, and an angle steering ring to the left front wheel. The success of this tractor led to the sale of one exactly similar (No.32029) to H.Lancaster & Co. of Newent, Gloucestershire, no vast distance from Pershore. Lancasters had it for about seven years but sold it in 1921 to Thomas Walker & Sons of Tewkesbury, who took out the superheater before selling it on.

Altogether eleven superheated tractors were sold in 1914. These included No.32360 for William Aitken of Linlithgow, who subsequently built up a fleet of Garrett overtype wagons; No32473, bought by E.W. Rudd; and a pair, Nos.32524 and 32525, on Colonial boilers, for the Crown agents for the Colonies. These last two were fitted for oil burning, but the equipment was not supplied by Garretts who were merely required to provide a sleeve through the water space between the boiler front and the inner firebox to accommodate the oil injection pipe. The Crown Agents made the tractors subject to inspection, both during construction and under brake test. In common with other engines supplied to them at that time the two were required to be fitted with Holden & Brooke 3mm 'NH' pattern hot water injectors.

The outbreak of the first World War on August 4, 1914, had had a profound effect upon Leiston Works, in common with almost every other part of Britain. Two of the most active directors, Frank (Junior) and Stephen Garrett, together with many of their workmen, were members of the Territorial Army and were called up at once. Carl Remy, the sales director, as a German national, was cut off in his home country whilst Emerich Schmach, who, though of Teutonic origin, regarded himself in every way as British, was exiled to Godalming, Surrey, where he had to live in a more or less isolated state, fearful to leave his house because of the risk of being taunted by bigots incited by such rabble-rousing publications as *John Bull.*

Amidst these external upheavals, the company, under the administration of Frank Garrett (Senior) and his second son, Alfred, continued to go through the motions of peace-time activity like a decapitated chicken. The export market in portables and threshers was in heavy decline but the demand for steam wagons increased whilst that for tractors held steady. The actual figure for the production of tractors in 1915 was forty compared with forty-one in 1914, but rollers declined from twenty-three to a mere three. Exports of tractors during 1915 came down to two only, the first of which was No.32556 made to Pilter's order and sent away at the end of March. This was one of the tractors actually built in the Autumn of 1914 and it was altered by having the steering and reverser transposed, with the tender cut-out on the right-hand side. The other export, despatched in May for a client of Chambers in New Zealand, was No.32685, built with a Colonial firebox and convertible to a roller. An Allen scarifier was sent with it, whilst in deference to the all-purpose role for which it was obviously intended it had governors. A feature of the tractors delivered in 1914 and 1915 was the increased popularity of the tubular water-heater, first fitted in July, 1914, to No.32360 delivered to William Aitken in Linlithgow. Seven water-heaters were fitted in 1914 followed by eight in 1915, but thereafter the numbers declined.

During 1916, despite the shortages and restrictions that resulted from the war, the production of tractors actually increased slightly so as to reach forty-two, though it dropped by ten during the following year. The Works was heavily engaged upon shell production and had lost many of its more vigorous workmen to the armed forces but because the demand for portable engines for export had declined so much it was possible to keep up the output of wagons and tractors. So far as civilian use was concerned the war had produced shortages both of horses and of the provender wherewith to feed them, thus turning attention increasingly towards the use of steam wagons and tractors. One would be hard put to analyse how the effects of the war had impinged upon the range of users. A number of tractors went to work on road maintenance or improvements. Thus Nos.32724 and 32725 were sold to Cork County Council and No.32923 to Cornwall County Council. Others, such as No.32752 for Tom Everton of Droitwich and No.32770 for the Oxford Steam Plough Co., were put to work in the haulage of roadstone. Others again were bought by firms involved in the building and construction industries either as contractors or as suppliers of materials. Austins Ltd of Ilford, Essex, who were woodworkers building sectional army huts, had No.32738; Parker Bros. (Mildenhall) Ltd. who were timber merchants, had No.32830, whilst Bowerman & Sons of Bridgwater, Somerset, bought No.32959 to haul bricks from their brickyard. A limited number went to the owners of large landed estates who were increasingly called upon to bring parkland and pleasure areas into cultivation and to fell woodlands to help fill the insatiable demand for pit timbers for the Western Front and the coal mines. Lord Rayleigh of Terling, Essex, acquired No.32782 whilst No.32784 was sold to R.W. Christy of Boyton Hall, Chelmsford. The Duchess of Hamilton had No.33019 at Easton Park, and the Earl of Ellesmere purchased No.33037 for his estate near Newmarket. A few tractors went to firms connected with farming and land work. One of these, No.33117 intended for the Rathbritt Cooperative Implement Society Ltd., was lost when the ship carrying it to Ireland was sunk by a German U-boat early in December, 1917. The majority of the remainder went to haulage contractors, most of whom would doubtless have been involved in war work of some description. Probably the best known of these was Norman E. Box of Ardwick, Manchester, who had superheated tractor No.33032 delivered to him in May, 1917. This was mounted on rubber tyres. Notwithstanding the fact that the country was at a critical point in the most disastrous conflict of the twentieth century Box insisted that the tractor be embellished to his requirements. Some of these, i.e. a drawbar specially designed to suit his business, check springs in the steering chains, an eye on the steering axle for attachment of a wire rope, and a tubular water heater, were sensible measures to fit the tractor for its work or to promote economy. It is more difficult, however, to see why he should have insisted on brass wheel caps and having the flywheel lined out, with particular care being taken to ensure that the lines ran true when the engine was turning over! The general scheme of painting was black, lined in vermilion and yellow to match his Fowler road locomotives, but it seems an unnecessary extravagance to have ordered the lettering to be carried out in gold leaf.

Change was in the air, however, concerning the supply of tractors for civilian use. The military authorities, belatedly becoming aware of the merits of mechanical traction, had begun commandeering tractors. A visit to Leiston Works in mid-April, 1917, by an Impressing Officer led to a number of finished, or nearly finished, engines being taken by the War Department. These included tractors Nos.33029 and 33034, ordered by E.W. & W.F. Crane of Great Fransham, Suffolk. The sequel to the taking of these two tractors was that the Ministry of Munitions entered into a contract [PM/MT 2728A(4) dated October 13, 1917] for thirty-two tractors for delivery by the end of May, 1918. These were of standard design, with an awning over the manstand. Sixteen of the early examples had 6" wide front wheels, probably using up wheels already made, but all others had 8" wheels. Only twenty-four were despatched during May, seven more straggling on into June and July, thus overlapping a second batch of twenty-eight ordered as an extension of the contract, delivery of which began in June and continued until October 1. By some sleight of hand of record keeping the last of the first batch was never made so that the total for this contract was only fifty-nine tractors. A second contract [29E/7351/C 1038] dated July 19, 1918, covered a further twenty tractors but the Armistice of November 11, 1918, led to the cancellation of the contract so that six of the tractors were actually sold to civilian customers. Twenty tractors had been asked for with Pickering governors and the inscription '70/ROADS' on the left-hand motion cover. Because the tractors were needed so urgently when the contract was let permission was given for the wheels to be constructed using pairs of angle rings to replace 'T' rings should this become necessary to obviate delay.

The interest of the Ministry of Munitions in Garretts' tractor output produced problems in supplying civilian customers. For example, to serve its isolated hospitals the Metropolitan Asylums Board had been in the process of replacing horse-drawn transport with steam tractors. They had already had Nos.32985 and 33078 when they found themselves obliged to apply to the Ministry of Munitions for a 'priority certificate' before a further tractor could be purchased. The issue of these certificates became fogged with bureaucratic inefficiency with the result that the authorisation for the third tractor was never forthcoming and in consequence it was not bought until after the end of the war. In the case of the fledgling DR type tractors described in Chapter 8, the Ministry of Munitions contracts for steam wagons and tractors, coupled with a contract for the building of aeroplanes, meant that the DRs could not be put into production at all until the war was over.

The War Department tractors were delivered by road in batches to the assembly point at Kempton Park, Surrey, in much the same way and by the same group of drivers as the War Department orders for overtype steam wagons. This was described in *Garrett Wagons* - Part 1. It was the good fortune of the delivery crews that the bulk of the tractor deliveries fell between April and October, 1918, when the weather was relatively good. The tractors were delivered in the closing stages of the war - some, indeed, after the Armistice had become effective - and consequently saw little military use. What little work they did put in can have had little bearing upon the outcome of the struggle.

Patriotic motives apart, the orders must have been of value to the Works, though had they been permitted to do so they could doubtless have sold an equivalent number to civilian customers. The inability to supply tractors and, for that matter, steam wagons, in 1917 and 1918 had an adverse effect on the relationships between Garretts and potential customers as some of the latter turned to other makers such as Clayton & Shuttleworth, who were still able to deliver both steam wagons and tractors. After the war the surplus War Department tractors, some virtually unused, were sold off by the Disposals Board. This, in itself, must have lessened the demand for new tractors at the Works. Of the 104 tractors made in 1918 only twenty-six were sold for civilian use so that a considerable backlog of tractor orders developed which provided a flush of work in this field during 1919. One of the tractors delivered soon after the Armistice on December 7, 1918, was convertible No.33422 for Archibald Maconochie of Cudham, Kent. There was some irony in this as he was a personal friend of the Avelings. So far as Leiston Works was concerned his name subsequently came to be somewhat soured by the part he played, along with the Aveling directors, in formulating the amalgamation proposals that attracted Col. Garrett into the scheme that finally emerged as Agricultural & General Engineers Ltd, (AGE), an ill-fated mis-match of disparate firms that pursued a grandiose and ultimately futile course until it fell in the opening months of 1932, bringing Richard Garrett & Sons Ltd. down with it. I have already related the story of this misadventure elsewhere, and it is sufficient to say here that when AGE took over in mid-year 1919 the order book for tractors and wagons was buoyant even though the output of portable engines and threshers had all but withered away.

Whilst it is not possible to mention individually all tractors supplied in the closing months of the old ownership, a few may be instanced. The good name they seemed to enjoy in Northern Ireland was reflected by repeat orders from Curry & Dodd, who had No.33332, and John Harkness & Co., (Nos.33414, 33419, and 33545), and by a new order from W.W. Kennedy & Co. (No.33412), all three firms being in Belfast. The troubles in Ireland were making it very difficult to deliver more than a few miles outside Belfast city limits but No.33633 was sold, and delivered, to Samuel Rea in Antrim.

Norman Box had a further tractor, No.33503, fitted with solid rubbers on the rear wheels, necessitating a new design of wheel to suit. In 1920 Box bought a third tractor, No.33713, this time superheated and again fitted with rubber tyres to the rear wheels, suggesting that the first two must have pleased him. On the strength of that supposition Victor Garrett approached him for, and obtained, a testimonial for use in sales literature, only to find not long before it was published that Box had given a similar recommendation in equally glowing terms to Fowlers. By this time superheating was coming under a cloud at Leiston. Only one further superheated tractor, No.34503, was sold and that was for export. As superheating went out the use of piston valves was dropped with it, although No.33434, a convertible sold to the estate of the Earl of Leicester at Holkham, Norfolk, was fitted with piston valves despite having no superheater.

Despite the fact that in pre-war days the firm had had little success in selling either tractors or rollers to local government bodies, in the brief euphoric period leading up to the onset of AGE control they managed to sell two to Lancashire County Council. The first of these was No.33426, which went at the end of March, 1919, to be followed by No.33459 in May. As the amalgamation approached there was a feeling of uneasiness about the Works. It was sensed that change would inevitably take place but few of the staff had any clear idea of what was likely to happen. What actually took place is the subject of Chapter 9, but before narrating those events it is necessary to step back a year or two and relate the course of another experiment, the designing of the 5NHP tractor for direct ploughing.

Fig.79. *The first of the DR tractors to be built was No.32974, completed and tested, still in primer, early in January, 1917. The photograph was taken on land not far from Leiston church.*

Fig.80. *No.32974 at work on Lord Rayleigh's farms in late Summer, 1917.*

Fig.81. *A front end view of the same tractor.*

Fig.82. *The third in the series of views on Lord Rayleigh's farms.*

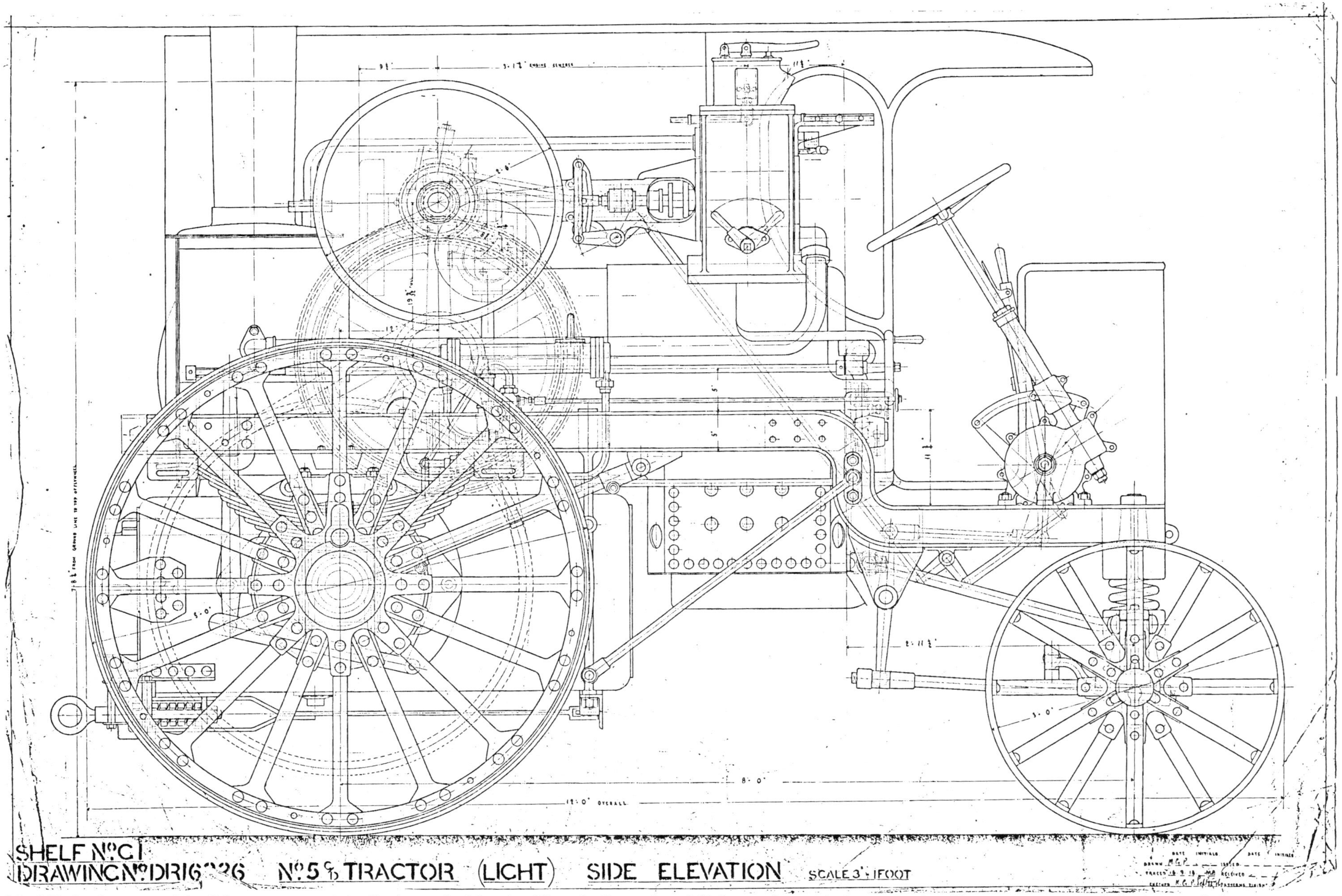

***Fig.83.** Offside elevation of the DR tractor as originally drawn by Maurice Plane in September, 1916, complete with the roof rather like a contemporary baker's cart. Only one tractor, No.32974, was built to this drawing (DR16326) and even that did not have the roof.* [S.R.O.]

Chapter 8

The DR Tractors

At the outset of the war the departure of Emerich Schmach from the leadership of the Leiston drawing office left a vacancy to which the board of directors elevated Henry Keay Pratt who had been on the staff since 1912. Pratt was a man characterised by Victor Garrett as of no mean engineering ability but taken up with concerns about his own status and importance. For the first eighteen months or so of his tenure of the post the two active directors were Frank Garrett, Senior, and Alfred Garrett, latterly supplemented by William Marshall. Pratt seems to have been on reasonably satisfactory terms with Frank Garrett, Senior, but when Frank, Junior, returned to civilian life in 1916 and resumed active participation in Works management his attitude to the chief draughtsman was much more critical. Possibly Pratt, basking in the approval of the senior and governing director, believed his position to be more secure than it actually was, but whatever the reason, his manner became increasingly irritating to Frank Garrett, Junior. Had he been a little more perceptive Pratt might have realised that in a contest of wills with the managing director the outcome could not possibly be favourable to him, but this seems not to have dawned upon him.

Development work on road vehicles lay in abeyance for the first twelve months of the war during which the drawing office was engaged upon setting up the shell production lines which involved designing lathes for the purpose to be made at Leiston Works. Vehicle development was revived mainly because the police forces of the country, for whom the Metropolitan Police became the mouthpiece, objected increasingly to the poor fields of vision afforded to the driver of an overtype steam wagon. With hindsight one is tempted to suggest that this could hardly have been a pressing matter at a crucial juncture in the worst war of this century but, nevertheless, it was pressed unrelentingly. In response to this stimulus the board issued instructions to the drawing office to investigate how this might be remedied, a task which Pratt delegated to Maurice Plane. The wartime embargo on development or experimental work not directly related to the war effort led to the objectives of Plane's project becoming two-fold, i.e. it was coupled to a proposal to build a light steam tractor for ploughing. For policy reasons work on the tractor was pushed a little ahead of that on the design of the wagon. So far as can be judged on the surviving evidence drawing work on the dual project began in the late Autumn of 1915. Plane's available courses of action were constrained by two factors. Firstly, the unhappy experience of the firm in 1904-08 when constructing the three early wagons (Nos.25065, 26721, and 26725) with undertype engines and vertical boilers had set the face of Frank Garrett, Senior, irrevocably against the use of the undertype arrangement which might otherwise have provided a solution to the field of vision problem. Secondly, because of the wartime demands on drawing office and Works staff, taken with problems of supplies of materials, he was encouraged, possibly even told point-blank, to design his revised wagon and the new tractor around existing components wherever possible.

As related in *Garrett Wagons* - Part 1, the resulting wagon, the XL type, used the principal parts of a superheated piston valved overtype but with the boiler reversed end to end so that the cylinders were over the firebox and the crankshaft at the smokebox end. The driver was at the very front thus giving him an admirable field of vision at the expense of a most awkward position of the throttle lever and a degree of personal discomfort unusually bad even in an era when the personal convenience of drivers was hardly considered.

When the same precepts came to be applied to a tractor they resulted in a machine of radically different type to the No.3 and No.4 tractors produced thereto which, in essence, had been scaled down traction engines in which the boiler acted as the chassis. In the new design of tractor, given the identification letters 'DR', this long-standing layout was set aside. Apart from the reversal of the boiler in common with the XL wagon, the components used in the boiler and engine were as in the general series of superheated 5 ton wagons, except that a revised superheater was fitted, though of course there were no hornplates required at the firebox end and there was a crank-box riveted to the barrel to carry the crankshaft bearings. The boiler was mounted in a chassis of 5" x 2½" rolled steel channel cranked down 11¾" at the leading end. The 3¾" driven axle was carried in axle boxes supported by 12 leaf laminated springs secured in shackles riveted to the frame. The fore and aft movement and chain tension was taken care of by radius/chain tensioner rods each side. Springing at the leading end was provided by a coiled spring in the central mounting. The tractor was two speed with the crankshaft gears outside the hornplates, or crank box, with the low speed nearest to the bearing. The fixed intermediate shaft was carried at its outer end on a bracket above the frame and at the opposite end in a mounting attached to the crank box. The fast and slow speed intermediate gears were fixed on either side of the chain sprocket in a bolted-up assembly. A roller chain took the drive from the driving sprocket to the driven sprocket and differential centre. The sprocket rim and the differential centre were each steel castings and the assembly was put together with mild steel spokes and fitted bolts, driven up. Via the nearside bevel gear, the two differential pinions drove the combined winding drum and driving boss which were free on the axle and from which the nearside wheel took its drive by a driving pin. Beneath the boiler the 3¾" load bearing axle ran in a cast steel cannon shaped hollow driven axle, upon the nearside end of which was fixed the other bevel gear. At the offside end of this hollow rotating axle there was a flanged and spigotted connection to a combined driving boss and brake drum (for the external contracting brake). This boss was turned where it passed through the axle box and where it carried the offside wheel, the wheel taking its drive from it by a driving pin. Both wheels were held in position by the conventional collar and cotter fastenings which were protected with cast caps, iron or brass to the customer's instructions.

The disposition of the gears and differential, probably more easily followed by reference to Fig.83 than by an extended written description, was necessitated both by the need to utilise the maximum number of wagon parts, and by the extent of the spring movement. Thus despite the fact that the proximity of the axle to the intermediate shaft might have suggested superficially that an all gear drive was feasible this was ruled out by the relative movement of the shafts and the wracking action of the chassis. The wheels themselves followed the profile of the conventional traction engine type but were of altogether lighter construction, the centre spiders being of cast steel to which the spokes were riveted, a method also used on the front wheels.

At the leading end the front axle was a cast steel I-beam with a central mounting point engaging with a bearing casting bolted to the front of the frame. The axle was stayed back from its ends to the frame and from the centre by a low level stay to the firebox outer plates, an arrangement that took care of wracking movement. The steering was Ackermann type. There was a sheet steel apron plate to the manstand which was placed above the front axle. Seats for the driver and mate were placed on either side of the firebox with the result that the horizontal lever controlling the throttle and the speed gear change lever, also horizontal, were roughly on a level with the driver's cheek-bone. As originally drawn the tractor was provided with a light overall roof.

These arrangements were worked out and consolidated onto finished drawings over a period of perhaps six or seven months. Specification No.200 for the wagon was dated May 9, 1916. As this summarised the whole of the drawings it could be said, with some degree of certainty, to represent a date when the design was complete. In practice, a number of manuscript modifications were made to the issued set of drawings as the work of building progressed through the shops. By contrast Specification No.201, covering the tractor version - referred to on the specification as 'No.5 C/D Tractor. Light' did not follow until August 11. This marked the point where the tractor drawings were in a state at which production might have been commenced, and two days later the specification was issued to the Engine Room Stores which housed the working sets of drawings ready for handing out to the men on the machines and benches as and when required. The first to be put in hand was No.32974, the erection of which commenced on September 27, 1916, by the gang headed by Walter Rogers and was finished in time to take part in a test run on January 10, 1917. As already remarked, the boiler did not differ materially from that fitted to the standard superheated 5-tonner, except that the tubes were given a slight upward inclination from front to rear, but the superheater had been redesigned so as to contain 16 coils of 3/8" bore tube each 7.16 feet long, i.e. a total of 114.14 feet, giving a heating surface of 11.11 square feet internally, and 19.57 square feet externally. It had been arranged that the first vehicle to be finished containing the new superheater, whether tractor or wagon, would be tested to see how the design worked, and Maurice Plane was nominated by Pratt to conduct this test. A loaded 5 ton wagon calculated to weigh about 13 tons all-up was the load, and the trial began at the Works gate with the superheat at about 400°F. About 4 minutes from the start this had risen to 500°F, and reached a maximum of 550°F, at which it flattened out, after 6.82 minutes and some 1200 yards from the beginning. This was considered to be, and indeed was, quite satisfactory, but the conditions were too limited and non-typical to reflect day-to-day use where the ingredients of success, namely the hot start, heavy load and unobstructed road were less likely to happen.

On January 23 the tractor was given a short practical trial at ploughing. H.K. Pratt's report to Col. Garrett was as follows:

Mr. Frank

re Ploughing Tractor

This was ready to run at 4 p.m. It took 8 furrows 6½" to 7" deep, speed 320 yards in 4 min. 35 sec. Width of ground turned 6 ft. = 2 yards. Square yards turned per hour = $\frac{320 \times 2 \times 60}{4.583}$ = 8370

Loss at headland - say 20%

Nett area ploughed per hour = 6950 sq. yds. = 1.44 acres.

H.K.P.

This note and its over optimistic result illustrates, perhaps, how Pratt's mannerisms contrived to irritate Col. Garrett so much. On the basis of four and a half minutes ploughing in the ebbing daylight of a winter day he had based an estimate of the work capacity of the tractor. In itself, this could have been tiresome, but to have deemed it necessary to point out that 6 feet equalled 2 yards was probably even more irritating.

Probably he was told to carry out further and better trials. These were arranged to be of greater duration and were spread over several days. The report to Col. Garrett, dated February 26, 1917, was unsigned but the draft is in Pratt's handwriting. This time the testing was more comprehensive but, nevertheless, was still not sufficiently long to give a thorough picture of the tractor's likely behaviour when in use on a farm or estate. The report, quoted below, clearly was drafted for use in advertising:

February 26, 1917.

SPECIAL STEAM AGRICULTURAL TRACTOR

Designed for ploughing, threshing, reaping, general farm work and road haulage.

Cost of Motor	£600		
Assumed life for purposes of sinking fund	8 years		
Sinking fund at 4% interest			
Number of working days per annum	300		
Hours per day - 10	Hours per year	-	3000

It should be quite possible, on a reasonable size of Farm, to keep such a machine fully employed.

Standing charges (Annual)

Interest 5% on £600	£30. 0. 0.
Sinking Fund - Annual amount to be set aside to give £600 in 8 years	86.16. 6
Insurance	16. 0. 0
Driver's wages at 35/- a week	91. 0. 0
	£223.16. 6

Particulars of trial when ploughing 6½" to 7" deep:

Duration of trial	1 hour, 20 minutes	
Coal used	84 lbs.	
Coal used per hour = 84 x 60/80	62 lbs.	
Area ploughed per hour = 1 x 60/80	0.75 acres	
Coal used	Nixon's Navigation at 42/5 per ton.	
Cost of coal per cwt.	25.45 pence	
Cost of coal per hour	14.1 pence	
Total cost per hour when ploughing 6½" to 7" deep:		
Standing charges	17.9 pence	
One man for ploughs at 35/- a week	7.0 pence	
Repairs at £40 per annum	3.2 pence	
Lubricants at £10 per annum	0.8 pence	28.9
Coal 62 lbs per hour at 42/5 a ton		14.1
Cost in pence per hour		43.0
Cost per acre = 43 ÷ 0.75 = 57.3d.	(say) 4/10.	

Particulars of trial when ploughing 8½" to 9" deep:

Duration of trial	2 hours 49 minutes = 2.817 hours
Coal used	272 lbs.
Water used	2166 lbs.
Coal used per hour 272 ÷ 2.817	96.5
Water used per hour 2166 ÷ 2.817	768 lbs.
Area ploughed per hour 2 ÷ 2.187	0.71 acre

Coal used	Nixon's Navigation at 42/5 a ton	
Cost of coal used per cwt.	25.45 pence	
Cost of coal used per hour	21.9 pence	
Total cost per hour when ploughing 8½" to 9" deep:		
Standing charges	17.9 pence	
One man for ploughs at 35/- a week	7.0 pence	
Repairs at £40 per annum	3.2 pence	
Lubricants at £10 per annum	0.8 pence	28.9
Coal 96.5 lbs. per hour at 42/5 a ton		21.9
Cost in pence per hour		50.8
Cost per acre = 50.8 ÷ 0.71 =	71.7 pence (say) 6/-	

In the above trials a 4 furrow plough was used as it was impossible to obtain delivery of a 6 furrow plough. By attaching additional ploughs to the 4 furrow one used, it was conclusively proved that the Steam Motor was capable of hauling 8 furrows when ploughing from 6½" to 7" deep. With a 6 furrow plough there is no doubt that the area ploughed per hour would exceed one acre.

In July, 1917, a further and better trial was set up at Foulness Island. At that time Foulness had no bridge to the mainland and getting the tractor there involved driving it across a causeway at low tide. Once on the island it was given the task of breaking up old compacted pasture land on a heavy soil that was to be brought into use as arable for the war effort. After a few trial runs had been made to enable the crew, which included Maurice Plane as the observer and reporter, to become accustomed to the conditions, on July 4 a trial was conducted on an area which had been selected and marked out so as to be exactly two acres, and was known not to have been ploughed for at least fifteen years. It proved to be very heavy work. Although the tractor worked fully spudded up to give it traction the dryness and compacted nature of the soil helped its performance by preventing it from sinking in. On the other side of the equation, however, the same soil properties made hard work for the plough, on several occasions bending the coulters. Ploughing began at 3.20 p.m. and ended at 6.33 p.m. during which time fifteen minutes had been spent changing bent coulters. Thus, after deducting this time loss, the nett period of ploughing was two minutes under three hours. The intention of the trial was that the depth of ploughing should never be less than 6" nor more than 9", and the plough used was a 4-furrow Ransomes, cutting a 9" furrow. During the trial the tractor used 384 lbs. of a poor quality steam coal, the B.Th.U. value of which was not recorded, and 2200 lbs. of water, i.e. 5.7 lbs of water were evaporated per pound of coal burned. This is the measure of the quality of the coal used as in an earlier trial, burning best Welsh coal, a figure of 8 lbs. had been obtained. Maurice Plane's evaluation of the data he had obtained was tabulated, avoiding the shortcomings of Pratt's earlier calculations:

	Per Hour
Standing charges, including depreciation, sinking fund, driver's wages at 35/- per week plus insurance	19.9 pence
Man on plough at 35/- per week	7.0 pence
Repairs at £40 per annum	3.2 pence
Lubricants at £10 per annum	.8 pence
129 lbs. of coal at 1s.7d per cwt.	21.9 pence
*Total cost per hour	50.8 pence

*Correctly this should have been 52.8 pence

Since two acres had been ploughed in 2 hours 58 minutes he worked out the cost per acre at 6s. 3½d. The trial, however, had begun with the tractor already at working pressure, and the costs of carting coal and water to it in the field were excluded.

Geoffrey Marreco, who looked after publicity, had in mind the projection of the tractor as an answer to the wartime shortage of horses that had resulted from Army demands to which end a satisfactory trial in the hands of an impartial user seemed a prerequisite. In September, 1917, the tractor was sent to work on Lord Rayleigh's farms in South Essex, managed by his agents, Strutt & Parker, at Witham, Terling, and Foulness Island.

During 1917 ploughing by tractors received extended notice in the trade presses of both the farming and implement trades. Numerous American motor tractors had been imported and tried, some, such as the International Titan 10/20, with successful results but others, of which the Bates Steel Mule may be cited as an example, with much less success, and, of course, there were also a number of British makers trying to find a way into the market. It was an article on one of these, the Pick Motor Tiller, appearing in *Country Life*, the magazine of the actual and aspiring landed gentry, in the Autumn of 1917 under the pseudonym *Ploughshare* that gave Geoffrey Marreco the chance to insinuate the Garrett DR tractor into the notice of the readers of the magazine. In the issue of January 26, 1918, under the signature *Tillage* but actually from Marreco, there appeared in its correspondence column the following :

SIR, - I am interested to read the article by 'Ploughshare' in a late issue dealing with the Pick Motor Tiller, and I hope that the following description of another purely British farm tractor will be of interest. This particular tractor has been working most successfully for some time on a big estate in East Anglia, and the results, to the writer's personal knowledge, have been most satisfactory, in spite of the fact that both the weather conditions and the nature of the soil have been far from ideal. The 'Suffolk Punch' Agricultural Steam Motor is somewhat of an innovation, seeing that, while primarily intended for ploughing, it can with advantage be used all the year round on the farm, and will not only drive a thresher and other agricultural machinery, but is also suitable for the haulage of farm produce to market or otherwise. With the present price of petrol the fact that any good, hard steam coal can be used appears worth consideration alone, and the fact that this motor can be kept in almost constant employment very much increases its profit earning capacity. The steam motor hauling a four-furrow plough made figures fluctuating from .70 to .75 of an acre on soil of varying character, at depths of from 6 ins. to 7 ins., the lower figure being on stiff old pasture, which made the ploughing very difficult. The cost varied from 4s.10d. to 5s.7d. per acre.

The fact that the machine is intended for continuous work all the year round and that it has a life very much in excess of the light internal combustion Agri-Motor, necessitates a certain increase in dead weight, but with the wide (14 in.) rims with which the driving wheels are supplied, the writer was particularly struck by the absence of undue pressure on the land or 'packing' of the soil. The engine appears to be quite simple in design, and is certainly easy to drive, so that any farm hand of average intelligence should quickly make himself efficient, and the driver's position in front, with motor car type steerage and all levers readily accessible, forms a distinct improvement on the ordinary direct ploughing steam tractors or road haulage engines. The makers (Richard Garrett & Sons, Limited, of Leiston) appear to have paid particular attention to the question of suspension, with the result that the motor closely conforms to the inequalities of the ground, even when ploughing on very uneven land. The general design can be clearly seen from the illustration, the engine being of the horizontal compound type mounted on the top of the boiler, which latter is fitted with a patent corrugated firebox, increasing materially the heating surface, and with a patent arrangement of superheater coils in the smokebox, resulting in a large increase in the expansive property of the steam, and consequent reduction in the fuel consumption. The drive is by machine-cut gearing for the first, and silent roller chain for the second transmission. The two-speed gear, giving road and ploughing speeds of five and two miles per hour respectively, is capable of adjustment up or down within reasonable limits, to suit individual requirements. - TILLAGE

The calculation of the price per acre of ploughing done with a DR tractor which Pratt had prepared for Col. Garrett and which Marreco had used in his press releases was flawed in several particulars, the most glaring of which was the assumption that a tractor would be used on 300 days of the year and 10 hours a day at that. After deducting Sundays and Christmas Day the working year was reduced to 312 days. To have worked it on 300 of these would have meant a utilisation factor in excess of 96%, a questionable proposition indeed, compounded by the bland assumption that on each of these 300 working days ten hours of useful work would be done. No allowance was made, moreover, for the cost of consumable stores such as gland packings and mud-hole rings nor for the cost of taking coal and water to the tractor in the field. There was the further consideration that, for the costings to be realistic, provision ought to have been made for the expense that would inevitably have been incurred in turning over by horse drawn plough the relatively large areas of field taken up by those parts of the headlands which the tractor could not negotiate.

On the design side the trials from January to March had disclosed one or two weaknesses that required attention although Maurice Plane left no detailed notes of what they were. These called for modifications as the trials proceeded. As in many projects of this type the modifications were on an *ad hoc* basis, and the alterations to the drawings represented the final conclusions rather than the steps on the way to them. The number of changes to the drawings was sufficient to warrant entirely rewriting the specification. Thus after a lengthy gestation period Specification No.202 came out on April 25, 1917, entitled 'No.5 C/D Agricultural Steam Motor with Superheater'. The ball and socket mounting of the front axle was found to be inadequate, and eventually it was redesigned on drawing DR 17054, dated August 1, 1917. Again the superheater coils were amended, though no date is available. Before starting at Terling new links, eccentrics, and balance weights 'similar to those on the wagons' were fitted. As the boiler had proved not to produce steam as rapidly as had been wished it was decided in future to fit the large, i.e. Colonial, size of firebox as standard and only fit the smaller box if specially ordered by the purchaser, so this involved making the frame longer. At the same time the pressure was slipped up from 200 p.s.i. to 220 p.s.i. Whether or not No.32974 had the increased pressure when working in Essex is not known. To assist the steering in difficult conditions demountable steering angles were provided on the front wheels and 6" wide bolt-on extension rings for the rears. For future production the 100 gallon tank capacity was stepped up to 187 gallons. In the drive train the crankshaft gears were dimensionally revised necessitating a slightly longer crankshaft, and the differential gears were made stronger.

At first No.32974 was named *Farmers Boy* but before long someone, probably Geoffrey Marreco, had the inspiration to call the type *The Suffolk Punch*, thereby setting it apart, it was hoped, from the motley variety of petrol and paraffin engined tractors on offer. Although No.32974 had been modified during the period of building and Works trial it is not certain that it was ever fully in line with Drawing List No.201. It was registered as BJ 3345 on March 12, 1917, having up to that date been run on trade plates when on the public road. The colour was described as 'service grey lined in white', which made it photograph well. Eventually it was sold, through the Irish Agricultural Wholesale Society, to S.J. Brown of Ard Caein, Naas, and sent off on May 25, 1918. In the haste to get it put on the train 50 yards of ½" wire rope were left on the drum instead of the 25 yards intended to be sent so that the purchaser received an unexpected bonus. The last reference to it in the books at Leiston is on January 24, 1931, when spares for it were invoiced to Brown.

The second example to be built and, of course, the first to follow the updated specification was No.33159 which had been ordered by the customer in May, 1917, but not delivered to him until June, 1918. In this instance the purchaser was much nearer home, namely F.W.D. Robinson of Roos Hall, Beccles. This tractor was last licensed in 1930 so enjoyed nearly thirteen years of use. The next in numerical order, No.33160, was not delivered until March 7, 1919, although the order had been placed on April 27, 1917, by the Southminster, Essex, office of Strutt & Parker on behalf of Lord Rayleigh. This tractor was also last licensed in 1930. No.33179 (February, 1919) and No.33180 (July, 1919) were both for East Anglian purchasers, the latter being bought by the celebrated John Goddard who farmed at nearby Tunstall, and both were late deliveries of orders placed in 1917. John Goddard, ever a realist, tried his tractor at ploughing after the harvest of 1919, found it to be dearer to use than his horses and never ploughed with it again. George Ewart Evans wrote in his book *Ask the Fellows Who Cut the Hay* that it was used for 20 acres and then scrapped. Actually it remained with John Goddard and his son until 1948 when it was bought for preservation by the late Chris Lambert. It is now in the collection of the Long Shop Museum. Leiston. On his return to Leiston Works after his war service, the late Jack Simpson rented a house in Tunstall from John Goddard. Jack, whose father was a farmer and knew the Goddards, came to be quite well acquainted with John Goddard and, because of Jack's connection with Leiston Works, the question of the tractor came into their discussions. Goddard had paid £772 for the tractor and was in no mood to throw it away. His reasons for ceasing to plough with it were more complex than Evans' reference suggests. The whole costs of working were, it is true, higher than horse work but in addition he considered too great a proportion of the fields was taken up by the headlands it left and which he had to plough with horses. He also found the weight excessive even when spread with the extension bands. Furthermore, a practical consideration was that the tractor driver, by reason of his position, could not readily see the plough. On belt work he found it governed badly but this fault disappeared when Jack dismantled and refitted the governor. It has been said that the tractor was awkward to set to the threshing drum but this was not so as drivers soon adopted the practice of facing the rear when setting, steering with their left hand and driving with the right.

What killed the chances of the *Suffolk Punch*, however, was the arrival on the market of the Fordson tractor in the closing eighteen months or so of the war. In Birmingham on company business in September, 1918, Denzil Lobley, one of the most gifted of the drawing office team, whose ambition was later to prove his undoing, wrote a letter of report to W.J. Marshall, the technical director. On September 23, from the Hen & Chickens Hotel, he reported on various matters, going on to say "For the edification of the agricultural community of the Works I am sending you some particulars of the famous Fordson tractors! Total weight of tractor with its complement of paraffin and water - 23cwt. (Think of this and compare it with the weight of our *Suffolk Punch*)". He set out calculations he had made which put the cost of ploughing with a Fordson at 5s. 6d. per acre, upon which he commented "These figures give me furiously to think as to the future, if any, of steam ploughing".

To my knowledge, at least three *Suffolk Punches*, ordered in the flush of enthusiasm generated by the publicity at the first appearance of No.32974 but not delivered because of the urgent order for No.4CD tractors for the Ministry of Munitions, were cancelled as a result. Larkworthy, the Garrett agent in Worcester, had ordered two (provisionally allocated Works Nos. 33172 and 33178) on May 7, 1917, whilst William Rawlings of Collingbourne Ducis had ordered one (provisionally numbered 33179) on

May 30, 1917. The latter cancelled the deal on November 21, 1918, and the former a fortnight later. These sales and probably others of which I am unaware were lost not simply as a consequence of the arrival of the Fordson, but also of the fact that the delay in delivery meant that the war-time incentive to increase arable acreage, which had triggered the orders in the first place, had been annulled by the Armistice of November 11, 1918. By the time the war ended over 6000 Fordson tractors had arrived in this country and the production of the Fordson went on to the point where, in 1921, it represented half the total number of American tractors made that year.

Nevertheless Lobley's prophetic words did not deter the putting in hand of further *Suffolk Punches.* Two more, Nos.33246 and 33247, were completed in September and October, 1919, and remained in the stockyard until March, 1920, when the first of the pair was sold through the Eastern Counties Farmers Cooperative Society to T. Francis of Toft Trees, Fakenham. The second one was used to fulfil part of an order for two placed by the London confirming house of Hornung & Co. Ltd. on behalf of the Sena Sugar Co. Ltd., for the completion of which order a further and, as it turned out, final tractor of this type, No.33508, was built. They left Leiston for Africa on March 20, 1920. Sena is on the River Zambesi in Mozambique, an area where steam ploughing, albeit mostly with cable tackle, was then becoming firmly established. No more *Suffolk Punches* were ordered but spares were still being supplied for these two ten years after they were delivered. The example supplied to Francis at Fakenham was noted in existence as late as 1945. All of the *Suffolk Punches*, therefore, had working lives in excess of ten years and at least two in excess of twenty years. Clearly it was not because of shortcomings in performance that the tractors did not succeed. Lobley had probably come near to the basic reason for their failure to sell when he made his reference to their weight and, by inference, bulk. Initial cost, too, must have had a lot to do with the matter. The price which started at £600 had gone up to £772 by the time the Goddards bought their tractor in 1919 and less than a year after that to £1025 each for the Sena Sugar Co. pair. None of this was helped by the exaggeratedly optimistic calculations of the cost per acre of ploughing with them that had been circulated. Whilst, admittedly, few farmers or estate managers were likely to have believed totally in any manufacturers' claims the disparity between what was claimed by Garretts and what could be achieved by an average operator was too great to be received well. The Fordson could undoubtedly plough at a lesser cost per acre than the *Suffolk Punch* and with less compaction of the ground, but on the belt the Punch would have out-performed it because it would have governed much better. Further, the greater weight of the Punch, though still much less than that of a traction engine, would have given it higher road adhesion and hence better control of a threshing machine or other hauled load on the road, but against these advantages there was the ever present fact that they were achieved at an initial cost over four times that of a Fordson. Having said all this, however, I remain of the opinion that had Garretts been able to deliver them in 1917 when the need was great, instead of perhaps twenty months later, the whole episode might have ended differently.

Fig.84. *No.33180, (July 15, 1919), owned by John Goddard & Son at Tunstall, a few miles from Leiston. It did less than a season's ploughing for them but remained in use for threshing and other belt work until the 1950s. It is preserved at the Long Shop Museum, Leiston.* [R.G.Pratt]

Fig.85. A wartime picture (1942) of Percy Cole's No.33566. [Allan Downs collection]

Fig.86. *No.33956, new on September 25, 1920, to George Chambers of Sissinghurst, Kent, passed from him to R.Peddle of Broughton, Hants, and thence in September, 1924, to R.H.Blunt of Minster Thanet, Kent, who owned it when this picture was taken.*

Fig.87. No.33289, one of the Ministry of Munitions 1918 tractors, was used after the war by Alf Dawson of Rushmere in whose hands it is seen here in Woodbridge Road, Ipswich, on Sept. 5, 1935.
[R.G.Pratt]

Figs.88 and 89.
For a short while after AGE was formed a logo was devised for the front of the tractors in which the 'big **G**' *was incorporated as the centre part of the AGE initials. At some point in its career No.33289 acquired a smokebox door thus equipped. At the time of these pictures the tractor belonged to S.&G.Whipps of Tolleshunt d'Arcy, Essex, who bought it in 1941 for use in wartime land reclamation.*

Fig.90. *Compound 8 ton three-shaft roller No.34788, supplied to the City Engineer, Dundee, Natal, in September, 1925, and still in use in 1959.*
[A.J.Martin collection]

Fig.91. *If anything could have saved the steam tractor it was the solid rubber tyre, but the cost was against it. Here Norman Box's No.32680 (March, 1915) was the pioneer as far as rear wheels were concerned, before AGE days.*

Fig.92. The first engine to be on solids all round was No.33828 (July, 1920), intended for furniture moving work. Here it is seen on a low floored wagon (in LNER and BR days known as a Lowmac*) at Leiston, ready for despatch.*
[SRO]

Fig.93. *No.34410 was turned out on full rubbers and in show finish in the Autumn of 1923 for the Smithfield Show and was bought early the next year by Thomas W. Ward Ltd. for their Silvertown depot.*

Fig.94. *No.34787 was started in July, 1925, shown at the Suffolk Show in June, 1926, and then had a long spell on the show circuit before being sold to Wards in July, 1927, (at a reduced price). The photo shows the last style of 4CD, with rounded corners to the fascia, a bead swaged into the motion covers, (and, not visible, into the gear case - see Fig.95).*
[A.J.Martin collection]

Fig.95. *Before delivery of No.34714 to J.Brace of Ongar, Essex, for timber haulage, the firm took some publicity photos at the Station Works in which, although it had no governors, it was paired up with various machines, here a maize sheller.*

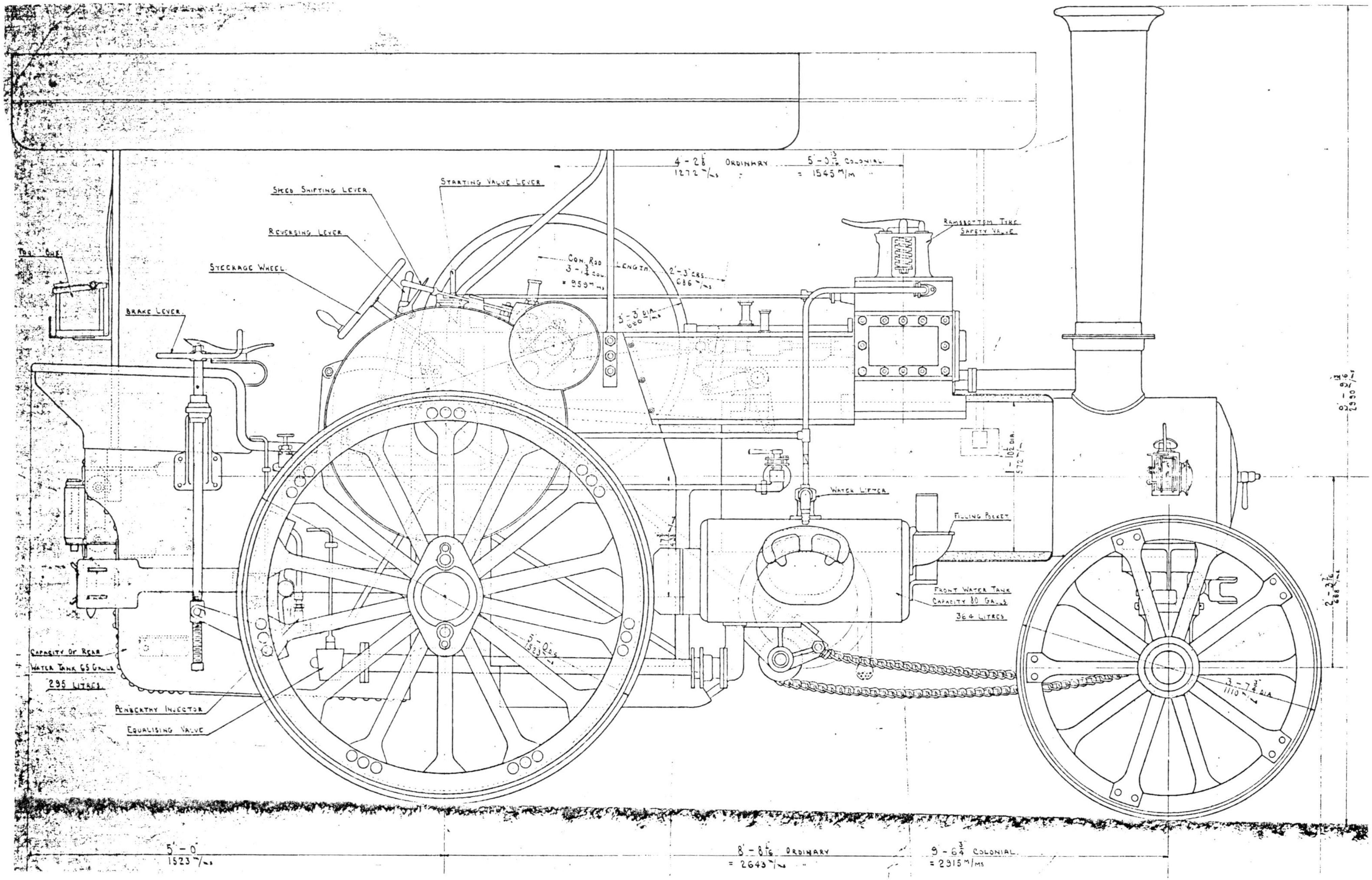

Fig.96. Gear-side view (Drawing No.CR 21031) of the No.4CD tractor as re-drawn in March, 1924.

Fig.97. Roller sales were in eclipse by AGE days but this is one of the three three-shaft compounds (Nos.34081, 34082, and 34083) that were sold to Japan via the London firm of E.H.Hunter in 1922. The characters in old Japanese on the motion covers translate roughly to The original ironpot, *obviously a nickname.*

Fig.98. *No.33818 was ordered by Walter Odam of Thorney, Peterborough, with various extras which included encased steerage, extended awning, gun-metal slide valves, and a smokebox attached by angle rings and bolts to enable a complete front-end change for conversion to a roller. It also had polished false covers to the front end of the cylinders and a liner to the chimney. When photographed on Oct.17 1938 it belonged to H.J.Holland & Sons of Dallinghoo, Suffolk, and the roller parts had been discarded.*

[R.G.Pratt]

Fig.99. *The last superheated tractor made was No.34503 (June, 1924) for South America to the order of Balfour Lyon & Co. of London. It had a Colonial firebox, larger belly tanks, 14" wide rear wheels, a spark arrestor, and an extended awning. It was supplied with the six* Arato *tracking trailers shown here.*

Figs.100 and 101. The swan song of Garrett heavy rollers was written not in Leiston but in Genoa with the Garrett rollers produced by Ansaldo under the licensing agreement. This compound, owned by Figari & Co., carrying the number 502 and the date 1925 was probably a 15-tonner. At least one Ansaldo-Garrett is known to survive (Fig.102)

Chapter 9

Under the AGE regime

The amalgamation of Richard Garrett & Sons Ltd. into the Agricultural & General Engineers combine (AGE) which came into effect from June 4, 1919, at a period when the back-log of demand for tractors was still being met, posed the group management with the problem of how to rationalise the various types of production as between the constituent companies. The output assigned to Leiston in the long term was, per week:

12 Portable steam engines
10 Electric vehicles
6 Steam wagons
1 CCS stationary steam plant

The exceptions made in the case of each of the firms was that [a] customer preference should be catered for; and [b] that existing stocks of parts could be used up. Aveling & Porter had been formally allocated steam tractor building but these exceptions meant that, in effect, it was divided between Leiston and Rochester. As a result a total of sixty-five tractors were made at Leiston in 1919, followed by forty-six in 1920, though, by contrast, in the period from January 1, 1921 to the end of March, 1923, no more than six were turned out. Only in the year April, 1924, to March, 1925, did the total again reach double figures and even then only fifteen.

As with all such decisions on rationalisation the result led to in-fighting between the constituent companies as well as resistance by lobbying and formation of cliques on the main board where the constituent companies were represented by the senior members of the erstwhile owning families - the patriarchs of the clans so to speak. Had the demand for new steam tractors continued unabated there is little doubt that the tacit approval of the work being shared between Rochester and Leiston would have continued also, but the abrupt decline in the rate of sales killed off tractor building at both of these Works. In part this came about as a consequence of the release onto the market of numbers of ex-military steam tractors which satisfied a considerable section of prospective purchasers but also the lure of the internal combustion engine diverted users away from steam tractors altogether.

The idea of Leiston continuing to build rollers when this work had been allocated to Rochester seems to have been anathema to the new administration in the opening days of the amalgamation. Thus an order received from the firm of Becker & Co. in The Hague for two rollers for shipment to Soerabaya Maschinenhandel in the Dutch East Indies was passed on to Aveling & Porter. The two rollers, A.& P. No.9138, an 8 ton single cylinder with Colonial firebox, and No.9243, a similar roller but of 6 ton nominal weight, were shipped in September and December, 1920, respectively. They were invoiced through Garretts, though whether or not badge engineering had caused them to carry the Garrett name is not known.

In 1918/19 so positive was the belief at Leiston, at least at middle management level, in the future of the steam tractor that a considerable amount of drawing office time was spent in improving and up-dating the tractor drawings. The total of drawing amendments on the standard tractor were sufficient to require an entirely new drawing list (No.251 dated October 20, 1919) to be made for it. The alterations it covered were spread over virtually the whole fabric of the tractor. Whilst the tractors built after the revisions were essentially the same as those before them the changes could fairly be described as widespread even if not radical in concept.

The new drawings fell into two main batches. Curiously those with the lower serial numbers, i.e. 17034 to 17043 inclusive, were dated between April 7 and 22, 1919, whilst drawings with the higher numbers, viz. 17235 to 17266 inclusive fell between October 3, 1918, and March 4, 1919. Nos.17545 and 17546 followed on May 9, 1919. Between them they left few parts entirely unchanged. Sections redrawn included the front axle and wheels, complete with steerage; the cylinder with its front and back covers, false covers, and the bored trunk guides; the drive gearing, differential and rear axle, together with the rear wheels; the crankshaft and the crankshaft pinions. These alterations were the most important. In addition there were, of course, many consequential changes to items such as brackets and not a few revisions of minor items such as lagging rings, the ashpan, and the firehole door. According to the engine memorandum books the first tractor built to the new drawing list No.259 was No.33941, put in hand on May 31, 1920, and despatched on August 27 to the Metropolitan Asylums Board, Dartford. On the other hand Nos.33949 and 33950, for Perth County Council, which left the Works on July 29 and August 23 respectively were noted as being built to the old specification (No.36). However, No.33952 for Thomas Bond of Kingston, near Taunton, Somerset, was to Specification 259 and so were subsequent tractors. Thus 53 tractors were built in this improved version, not a negligible number in the circumstances though doubtless far short of what had been hoped for when the revisions were first put in hand.

The superheater arrangements were redesigned on drawing BCR17996, probably in an effort to bring about some degree of standardisation. This new layout used the headers already employed in the 5 ton overtype wagons coupled with the same 5/8" outside diameter tubes and supporting structure as were fitted in the DR tractors, eight banks on each side of the central space left for the exhaust pipe. This gave 114.6 linear feet of tube resulting in an external heating surface of 19.57 square feet. The final print of the drawing is dated March 2, 1920, but there is a note written on the negative which says 'This drawing used by erectors for some considerable time...' so that one may reasonably suppose it to have been used in sketch form ahead of its final version. The only superheated tractor built after March, 1920, was No.34063, noticed later on in this chapter.

A glance through the list of purchasers of Garrett tractors during the six months following the birth of AGE illustrates the good name and reputation enjoyed by the tractors and the reason why there could be no instant telescoping of the Leiston order list into that of Rochester Works. Tom Everton of Droitwich, a loyal Garrett customer, though mostly for wagons, made his final purchase when he had No.33526 a week after the merger. Before the month of June was out Harkness of Belfast had had No.33545, and Samuel Rea of Antrim, No.33633. Early in 1920 Norman Box took delivery of superheated tractor No.33713 *Trojan*, his fifth and, as it turned out, final Garrett tractor. To suggest what made him stop at five can be nothing more than conjecture but with his ability to think probably two moves further ahead than his transport rivals it seems to me more than likely that he decided to concentrate on the heavy haulage side of the business where effective competition was much less rather than to continue in the medium heavy field where the tractors were most used and where competitors were becoming thick on the ground. Further, as the late Tom McTaggart suggested, it seems probable that he saw motor vehicles taking a larger part in this type of haulage.

Others, not seeing so far ahead, continued with tractors. Down County Council, in the newly separated territory of Northern Ireland, faced with the double task of making good the damage left from 'The Troubles' and making road improvements, ordered four tractors and twelve 6-ton trailers. Delivery began in March, 1920, with tractor No.33799 and went on until the last of the trailer wagons was sent on June 26. The three other tractors were Nos.33800, 33801, and 33802 whilst the trailers were numbered from 15916 to 15921 and 15929 to 15934, in each case inclusive. In Scotland Perthshire County Council placed an order for two tractors (Nos.33949 and 33950), a 5 ton overtype wagon (No.33951), three 3-man living vans, and four 6-ton trailers. These, however, were not sent until July and August, 1920.

The order from the Eddison Steam Rolling Co. Ltd. of Dorchester, headed by the notorious toper, Joseph de Mattos, for four tractors and two trailers also reflected, albeit obliquely, local authority activity in that they were bought for the contract cartage of roadstone for Wiltshire County Council which preferred to hire rather than to own its plant. From first to last Eddisons never owned a Garrett roller even second-hand, de Mattos being a crony of Harry Pitts, who ran Aveling & Porter's sales department. By one of life's ironies, however, in 1926 Aveling & Porter, alarmed by the scale of the debts owed them by de Mattos, secured his withdrawal from the management of Eddison and the appointment of Victor Garrett in his place, charged with the task of pulling the company back from threatening insolvency. This he did with great success, leaving one wondering what might have happened had he, rather than his brother Frank, been the chairman of Richard Garrett & Sons Ltd.

Whereas scarcely a Garrett tractor had been sold to a local authority prior to 1914, in addition to the six noted above a further nine were sold to councils and corporations in 1919 and 1920. One, No.33706, was for Cardiff but the other eight, Nos.33632, 33646, 33665, 33671, 33698, 33838, 33953, and 33989 were for rural district councils, most of whom before 1914 probably had nothing more adventurous than a horse and cart. The change was not so much a reflection of desire on the part of local rate-payers to finance improvements as a consequence of the passing of the Ministry of Transport Act 1919. In the new

Ministry the personnel of the roads section was mainly that of the old Road Board moved en bloc, headed by Henry Maybury, by then knighted for the massive service he had performed for the Allied cause during the war in co-ordinating and improving the repair of vital roads close to the fighting lines. On Maybury's advice a system of grants was set up to enable small road authorities to effect road improvements and to buy the necessary plant. This it was that led to the flow of tractor orders from minor road authorities. As with most initiatives of its kind the initial flush of enthusiasm and fiscal generosity did not last very long, but even so Garretts still reaped from it a few more orders for tractors.

Spending on road plant by minor authorities continued, in fact, into 1921 before it lost its impetus. Martley RDC in Worcestershire bought No.33991 in August, 1921. Later the same month East Westmorland RDC had No.34008. Not all were bought by rural district councils, and orders were secured from larger fry. Early in March, 1921, Roxburgh County Council bought No.34019 for use in its Hawick District. After these three there was a pause until November, 1923, when convertible No.34330 was sold to Banff County Council whereafter the only sales to local authorities were in Ireland. Two convertibles, Nos.34484 and 34541 went to Waterford County Council and Kilkenny County Council respectively in July, 1924, and in the same month two straightforward tractors were sold, No.34517 to Roscommon County Council and No.34584, which was equipped with a steam take-off valve to provide steam for a rock drill, presumably for stone quarrying, to Tipperary County Council. All four transactions were the result of a sales visit by Victor Garrett to the Republic where roads had not only the same need of improvement as elsewhere but in addition had been ravaged by destructive action and lack of repairs during the Troubles and by the civil war which had followed them. The same visit by Victor Garrett produced roller sales as well which will be touched upon shortly. The final sale of a Garrett tractor in Ireland took place when No.34703 was shipped to Roscommon County Council in November, 1927. Again it was expected to serve more than one purpose, having a ¾" valve on the boiler to supply steam to jack hammers. It also had a Pickering governor as it was intended to drive a Goodwyn Barsby stonebreaker.

The period immediately under review encompassed the building of the last purpose-built showman's tractor, the celebrated *Medina* (No.33902) made for James Humphreys of Pelsall, Staffordshire. Apart from having the expected extended smokebox, dynamo saddle and shield plate with a full length awning carried on brass sheathed columns, it carried brass rings and stars so elaborate as to merit a special drawing (CR10418). This brasswork included star centres on the cylinder covers, a ring on the flywheel, stars and rings on the motion guards, and the fore tank and axle caps. The brake and steering wheels were brass over iron. With an eye on soft tobers it was ordered with 14" wide rear wheels and 7" wide fronts. The colour scheme, too, was elaborate. The colour of the engine and wheels generally was crimson lake, the wheels lined with gold leaf and black, and the engine itself in gold leaf, black, red and white. The underside of the awning was a brighter red. Its final fairground owner was Charles Presland, the South Essex showman, and it is now in the Thursford collection.

By the end of 1921 tractor sales were in steep decline with a total of only thirty-two sold thereafter, of which several were in stock or in Works use for long periods. Although none of these were for showmen it was in the twenties that the 4CD became popular with travellers. Many owners of side stuff had used horses until the war and many right through it, but when tractors and lorries came onto the market through War Stores Disposals they took the opportunity to mechanise, many seizing the chance offered to them by a steam tractor to abandon kerosene flares in favour of electric light. Showmen stuck to steam for generating for more than a decade longer - some for twenty to twenty-five years more - because it governed well, but during the thirties considerable numbers of steam tractors gave place to second-hand Tilling-Stevens petrol electric buses. Few big fairs of the 1930s could be considered complete without a line-up of them, the tops of the radiators open, all boiling merrily.

Although such tractors came from the disposal sales on steel wheels it was not long before examples in showland use began to appear fitted with bolt-on solid rubber tyres, with resultant improvement to driver comfort on the road and an increase in the gear ratios because of the 7" or 8" increase in the diameter of the driving wheels. These tractors were very frequently in the ownership of men who did their own driving and therefore appreciated the better riding qualities when on rubbers but the fitting of these was made more urgent by the coming into force of the Road Vehicles (Showman's Tractors and Heavy Locomotives) Regulations, 1927, which enshrined a concessionary rate of Road Fund taxation for tractors and road locomotives mounted on solid and continuous rubber tyres. One of the first, if not the very first, showmen's tractors to be inspected and approved under the new regulations was Charles Thurston's *Prince* (No.33111), a tractor ordered in March, 1917, but because of the Ministry of Munitions orders, not delivered to its purchaser, Benjamin W.R. Hillen of Woodham Mortimer Lodge, Maldon, Essex, until October, 1918. Hillen had it only for a year or two and by August, 1922, it was travelling with Charles Thurston. Later it was to belong to Edward Penfold of Merton, Surrey.

During the early twenties the firm continued to exhibit tractors at important shows. Thus No.34134 attended the 1922 Royal Show at Cambridge, from which it was sold by Teddy Ballam to Henry Hensby of Wickham Brook, Newmarket. Incidentally, this was virtually the last sale credited to Ballam, who, as head of the home sales office until the advent of AGE, had been responsible for selling hundreds of engines either by his own efforts or via the agency system he had devised and set up. Selling had been centralised as a group function by AGE with conspicuous lack of success, established customers making it clear that they wished to continue to deal, as Hensby had done, with the Leiston sales office and personnel whom they had known for years. Central selling finally ended in October, 1922, as Ballam approached retirement.

As noted in chapter 7, the fitting of solid rubbers to the rear wheels of tractors began in 1917 with No.33032 for Norman Box, but the difficulty of getting tyres during the war meant that no other tractors were so equipped until No.33258, for J. Hill of Manchester, came out in February, 1919. In May, 1919, Norman Box had No.33503 rubbered at the rear, and in the next twelve months seven tractors, Nos.33632, 33646, 33665, 33713, 33745, 33793, and 33795, were thus fitted. The first customers to take the plunge of having rubbers all round were J. Mellor & Sons of the Palatial Depository, Guide Bridge, Manchester, whose No.33828 was despatched on July 28, 1920. Two days later this was followed by No.33881 for the famous building contractors, Holloway Bros. (London) Ltd. of Grosvenor Road, Westminster. Surprisingly only two further tractors, Nos.34410 and 34787, were supplied fully rubbered. Both were owned by Thomas W. Ward Ltd. and worked from their Silvertown, London, depot.

No.34787 was put in hand on July 9, 1925, with Sid Poynter as the chargehand erector. Although on rubbers all round (1070 x 160 for 871 rim at the front and twin 1530 x 140 for 4' 6⁵⁄₁₆" rim at rear) it seems to have been built for stock. Nothing is recorded of its progress until its appearance at the 1926 Suffolk Show held at Stowmarket on June 3 and 4. From there it went to the Norfolk Show at Yarmouth, and on to the Highland at Kelso on June 29 - July 2. Probably because of timing it did not go to the Royal, held that year at Reading on July 6 - 10, and No.34789, destined for T. Parker & Son of Kingsdon, Somerset, went in its stead. However, No.34787 was on parade for the Smithfield Show from December 6 to 10, and was still available for both the Suffolk and Norfolk Shows in June, 1927. The wallflower at last found a suitor and on July 9 was sent to Wards, albeit at a somewhat reduced price of £787-10-0 compared with the AEA price of £992.

At a cost of £113, putting a tractor onto solid tyres was an expensive exercise and two clients attempted to achieve a similar result with lesser expenditure. When No.35129 was supplied to A. & W. Cushion in Norwich in 1928 it was ordered with rubber cross strakes. These were cut from salvaged lorry tyres bolted on in the same pattern as steel strakes. As they were seldom of exactly the same thickness and, in any event, were not continuous, they were not altogether satisfactory, giving a droning sound at any speed and an uncomfortably jarring ride. Cushions' engine was intended mainly for timber haulage and this may not have mattered to them all that much. I never knew either the owners or the engine and therefore cannot add any useful comments. I did, however, know the late Frank Seaton who drove the Traylens' Foster showmans engine *Queen Mary* when it was similarly shod. In reply to a question as to what sort of ride it gave he answered 'Like a b....y cog wheel'! Before delivery to the Cushions No.35129 was exhibited on the company stands at both the Suffolk and Norfolk Shows of 1928, thus having the somewhat dubious distinction of being the last 4CD tractor to appear at any show. The very last tractor sold by Garretts, No.35225 (September, 1929), was ordered by the customer, W. Stanford of Preston, Yeovil, to be fitted with bolt-on circumferential rubbers cut from old solid tyres. Both these and the earlier rubber cross strakes seem somewhat inglorious expedients to have been employed on tractors ex-Works. One wishes that the line could have come to an end in rather more dignified footwear.

Of the final thirty-four tractors actually built between January, 1921, and the sale of the last one in September, 1929, fifteen came into the firm's financial year from April, 1924, to March, 1925, as already noted, leaving a very thin spread over the remaining years. Of this post-1920 output fifteen remained in

East Anglia, a reminder of the good name enjoyed by Garretts in their home territory, even after the formation of AGE. Despite the fact that road locomotives and tractors were beginning to come upon lean times, as evinced by the cheese-paring on rubber tyres noted above, there were purchasers who continued to ask for the highest standard of finish. When J. Brace & Sons of the High Ongar Sawmills in Essex had No.34714 in February, 1926, they asked for it to be turned out in show finish, choosing a brown livery instead of standard green, with red wheels and brass wheel caps. It was stipulated that the rear wheels should be 14" wide with 'wide' strakes, and a 7/8" x 2" steel tyre was asked for over the standard tyre on the front wheels. This standard of finish was not sought for a tractor used, like No.33828, for moving the expensive contents of large houses over paved roads but for hauling round timber out of woods and fields and back to the mill. It was superseded on this work by a Foden, after which it was used on the fairgrounds, first by Charles Presland of Tilbury and, after him, by Job Martin of Belvedere, Kent.

Export selling, at which AGE head office had rather greater success, continued to be a central function but, as far as Leiston Works was concerned, had never recovered in any degree from the 1914-18 war, and what recovery did accrue was largely undone by the decision of the Baldwin government (brought to power at the 1924 election) in the person of its Chancellor of the Exchequer, Winston Churchill, to bring the country's currency back onto the Gold Standard and to re-establish the pound sterling at its old parity with the dollar, viz. 4.86 dollars to the pound. This made British exports, particularly coal and engineering products, overpriced for the world's markets. These two factors, one national, the other internal to Garretts, in combination meant that only three tractors were sold abroad during the period under discussion. The earliest of the three was No.34063, commissioned in 1920 for export to South America by a purchaser named Carolus Brown. Brown was represented by his agents Balfour Lyon & Co. Ltd., then at Eldon Street, London E.C.2 Ordered with a fuel rack and a Colonial firebox to enable it to burn wood, and with parts to make it convertible to a roller it was accompanied by six trailer wagons. The extra length of the firebox, taken with the fact that it had a full length awning, made it appear an exceptionally long engine. Other deviations from standard were the fitting of a geared pump, 6" wide front wheels, and a stove pipe chimney. The rear wheels were 14" wide which was greater than the nominal standard width but, in practice, was used for virtually all of this last group of engines. Equally interesting was No.34503 shipped to Chile on June 13, 1924, also to the order of Balfour Lyon & Co., by then of 5 Chancery Lane, London. This, too, had a Colonial firebox but it was also superheated and fitted with a spark arrester in the base of the chimney, and, again, the 14" wide rear wheels were used. To go with it Garretts supplied six of the *Arato* type tracking trailers, for which they had acquired the United Kingdom patent rights. A train of *Arato* trailers would follow faithfully along the wheel tracks of the tractor at its head, but there were compensating disadvantages, notably a tendency for the trailers to snake as the speed was checked when running downhill.

The last of the three export tractors was No.34855 for Ukanma Plantations Ltd. whose postal address was Chania Bridge, Thika, Kenya. They had been introduced to Leiston by a Mr. G. Stewart Watt. As in the two earlier tractors a Colonial firebox was used, a wood rack was ordered on the tender and a wire spark arrester for the stove pipe chimney. Once again, the awning was full length. Unlike the two previous exports, however, it was ordered with no governor, demonstrating that it was not intended for belt work, but rather for direct traction ploughing for which, like the New Zealand tractors shipped earlier, it had a double drawbar and 6" wide extension tyres to the rear wheels.

With road rollers the case was somewhat different in that output had already come down from a yearly average of 27.5 over the ten year period from 1905 to 1914 inclusive to 3.2 for the five years from 1915 to 1919 inclusive. It recovered to twelve in 1920 and to fifteen in the twelve months from April, 1922 to March, 1923, but the whole production of 43 rollers in the AGE period only surpassed by five units the output for the pre-war year 1912, when 38 were built. A survey of what was made is, perhaps, best begun by mentioning the surprising way in which the old *heavy* design of roller retained its appeal. In the concluding days of the family regime three rollers, Nos.33088, 33089, and 33090, were put in hand for the Commission Internationale de Revitaillement, which had an office in India house, Kingsway, London. These three were ordered as 12½ ton compounds, and the order was executed by supplying 12 ton (nominal) four-shaft compounds (code-word *President*), the actual weight of which was, of course, well over the 14½ ton mark. The first was finished and sent off in September, 1917, but the second and third ran foul of the War Department orders for steam wagons and tractors with the result that they did not go until November, 1918.

Quite the most remarkable thing, however, about the sales of rollers in 1919 and 1920, particularly in view of group policy, was the series of orders for four-shaft rollers placed by the London office of John Birch for shipment to Bombay. This was an area to which Burn, situated as he was at Howrah in the east of India, seemed to have little access. In 1919 Leiston found itself with orders for a dozen rollers for clients of Birch, all to be consigned to Bombay, the first of which, 10 ton 'heavy' single No.33631, was sent from the Works on September 16, 1919. The remainder followed at intervals through 1920 although the numerical order often did not tally with the date order. Two, No.33731, a *Reading* four-shaft 10-tonner, and No.33732, a compound four-shaft 10-ton *Autocrat*, left on March 10. In July, 5 ton three-shaft single No.33808 was sent, having been preceded in May by the next in numerical order, No.33809, a three-shaft single sold as a 7½-tonner. Both of these had cast steel rims and forks. The next was No.33815, a four-shaft heavy 10 ton single sent on August 12. In this instance the consignees are known to have been Turner Hoare & Co., Engineers, Bombay. A trio of four-shaft rollers (code-name *Ruler*), Nos.33856, 33862, and 33863, were sold as '9 tons empty', the first being despatched on July 15, and the other two on July 26, but No.33857, a *Baron* three-shaft light 10-tonner, and No.33859, a 12 ton compound, had already left on June 24 and No.33858, an *Autocrat* four-shaft heavy compound, sold as weighing 12¾ tons, went at the same time as No.33856 on July 15. The building of this batch of rollers was somewhat upset by a strike of the moulders in the foundry. The promised delivery date of No.33815, for instance, was set down in the order book as 'Six weeks after the end of the moulders' strike'.

After the completion of the Birch orders there was a pause of eighteen months or so, until three *Barons*, Nos.34081, 34082, and 34083, were sold via E.H. Hunter for service in Yokohama and Kobe, Japan. The first of these had been built for stock and used briefly on the Works, being licensed as BJ 6965, before being refurbished and exported. They went away in February, March, and June, 1922. Since nothing defeated a Garrett signwriter they carried lettering in the Japanese alphabet. Two other rollers, both *Barons*, were later shipped to Bombay to the orders of purchasing agents other than John Birch. MacBeth, Taylor & Co. ordered No.34205 (October 10, 1922) and Robert Taylor & Co. Ltd., Victoria Street, Westminster, ordered No.34266, which left on November 24, 1923. Concurrently with No.34081 two *Barons* (Nos.34084 and 34085) were delivered to East Suffolk County Council taking the fleet numbers 3 and 4. Later they had No.34267 which became their No.5. In September, 1922, yet another *Baron* (No.34165) was delivered to Zachariah Fairclough, the man who had developed so much of Clacton-on-Sea. He named it *Clacton Queen* after the paddle steamer that, in its own way, had also contributed to the growth of Clacton. Like many seaside queens the roller had a life of steady decline. From Fairclough it passed to A.M. Emery of Hadleigh, Suffolk, and in 1953 from him to John Gray, also of Hadleigh. At the latter's sale in 1964 it became part of the Philp collection at Castle Hedingham but, unfortunately, was allowed to decay. When the collection came up for sale in 1995 the roller was bought by Tendring District Council who leased it to the Clacton Queen Steam Preservation Society, formed to restore it, a task that was achieved in 1997. Fairclough had a second *Baron*, No.34706, at the end of March, 1925, naming it *London Belle* after another well known paddle steamer. This had a similar history to *Clacton Queen* save that at the Philp sale it was purchased by a buyer from the north of England whose stated objective was to convert it to a tractor.

On the waning export front two *Barons* went to Thomas Cook at Port Said; No.34166 in August, 1922, and No.34204 in November, 1922. Two further examples, Nos.34173 and 34174, left for Karachi in September, 1922, via the agency of Wm. Menzies. Only two more *Barons* remain to be mentioned, both exported to Spain - No.34217 (January, 1923) to La Casa Metzger S.A. in Barcelona and, a month later, No.34265 to Julio Peña y Martin, Madrid. Whilst the first has gone, the second was found out of use at Ibiza in 1974. In 1976 it was presented to the Transport Trust and was repatriated, and now forms part of the collection at the Long Shop Museum, Leiston. There were two other sales of three-shaft rollers, one of which, the 4-tonner No.34960 exported to Rangoon, has already been mentioned in Chapter 5. The second, No.34788 for Dundee Municipality in Natal, was an 8 ton *Prince* mounted on a Colonial boiler, sold through Hubert Davies in Johannesburg, whose main interest in seeking the Garrett agency had been the undertype wagons.

The remaining twelve rollers to be described included seven four-shaft singles sold by Victor Garrett during his visit to Ireland early in 1924. He was introduced to a group of county surveyors by the brothers R.F. Bowen, county surveyor of Kilkenny, and J. Bowen, county surveyor of Waterford. Whilst his brief was not to sell Garrett rollers, he had established a rapport with the surveyors and finding himself in the situation where he could sell a Garrett roller Victor thought (to quote his own words to me)

'To hell with it' and captured the orders for Leiston. One more four-shaft single, No.34253, a *Reading*, was sold in the British Isles, bought by the Borough of Newport, Isle of Wight, in March, 1923, and scrapped in 1953 after an exemplary but uneventful life.

Roller building at Leiston ended, as it had begun thirty years before, with a pair of four-shaft heavy type rollers for a foreign buyer. Both were 10 ton compound *Autocrats*, ordered by the London firm of Gellatley Hankey & Co. for use in Arabia. No.34704 left the Works on October 3, 1927, and its sister (or brother?) engine No.35131 went away on January 17, 1928, the very last to go. The year 1927, therefore, appeared to mark the end of Garretts' involvement in roller building and so it did in the direct sense, but Garrett rollers clung on to life for a few further years as a result of negotiations carried on with Ansaldo S.A. the Italian engineering conglomerate. Despite having a relatively large engineering industry Italy had only one long-standing builder of road rollers, Puricelli of Milan. Under Mussolini's regime a move was undertaken to reduce the degree of Italy's dependence upon manufactured imports, and it was in accordance with this policy that Ansaldo began to be interested in roller manufacture, though whether it was as a consequence of ordinary commercial assessment or direct prodding by the Italian government is not known, and at this stage seems unlikely to be known. How this interest, once established, came to be linked with Garretts likewise remains a mystery. Leiston Works had supplied sundry rollers to Italian purchasers before the first war, including a pair to Genoa where Ansaldo proposed to build their rollers, so that the chance sighting of these at work may have prompted an approach in 1924 from Ansaldo to Garretts. Ansaldo was represented in London by Graziani & Co. Ltd. who had an office in Westminster. Graziani was related to Count Graziani who held high ministerial office under Mussolini. Once the contact had been made Victor Garrett was deputed to act for Leiston in dealing with Graziani.

What was requested was a licence to build the Garrett range of rollers at Genoa. In the narrowest sense Leiston had nothing to lose by granting such a request as with group policy being to concentrate such roller orders at Rochester, Garrett roller sales any way were at their last gasp. In the context of the AGE group as a whole there was perhaps rather more to lose as each roller built in Italy meant at least one less potential chance for Rochester to obtain an order. How or why Ansaldo picked upon the Garrett designs when more modern types existed at Rochester is also a little puzzling. The designs of forecarriage head and swivelling forks were, it is true, excellent and superior to those in use at Rochester, whilst the design of the cylinder block, engine work and gearing had earned itself a good reputation, particularly in the rollers based upon the 4CD tractors. These factors may have influenced Ansaldo's choice but again it may all have been chance. Victor described Graziani to me as a suave and gentlemanly, but shrewd, negotiator. He knew what he wanted to achieve and knew how to do so. Graziani really held all the aces in this matter. He let it be known, albeit in a polite and genial manner, that in the event of a licence to build not being granted at what his principals considered reasonable rates it was likely that they would build unlicensed versions of the designs secure in the knowledge that the complications of the Italian civil courts and the obstructive tactics of the Italian government would ensure that any legal retaliation by Garretts would be blocked. Faced with this situation Victor might justifiably have felt that his ace had been trumped had he not, for his part, considered the designs by that time financially worthless to Leiston so that any payments he negotiated with Ansaldo would represent pure gain. Agreement having been reached and a contract brought into being, sets of drawings were prepared for Ansaldo, together with drawing lists as set out below:

334	07/10/24	12 & 15 Ton c/d Road Roller
339	14/01/25	8 &10 Ton c/d Light Road Roller
345	07/02/25	10 Ton c/d Road Roller
351	31/03/25	18 & 21 Ton c/d Road Roller
363	18/09/25	6 & 8 Ton s/c Road Roller
368	26/10/25	5 Ton s/c Road Roller
370	06/11/25	4 Ton s/c Road Roller
631		10 & 12 Ton s/c s/h Road Rollers
632		8 & 10 Ton c/d (Light) Roller with Superheater

What we do not know is how many Ansaldo-Garrett rollers were produced since I have been unable to trace in the surviving Garrett account books any reference to monies received from Ansaldo. Possibly the sums received were paid through the group head office in Aldwych. Certainly at least two four-shaft examples were built. One was used by the Genoese contracting firm of Figari & Co. (Figs.100 and 101). At the time of writing the other one (Fig.102) survived in a public park at Milan.

Fig.102. *The known survivor of the Ansaldo Garrett rollers in a play area near Milan.*
[P.Gulliver]

Fig.103. *Tractor No.34561 (Sept. 1924) was new to S.R.Sherwood of Playford, Ipswich, but in 1928 was added to the fleet of H.J.Holland & Sons of Dallinghoo, Suffolk, one of the firms who made very successful use of their tractors for threshing. The picture was taken at Pettistree, near Wickham Market.* [R.G.Pratt]

Fig.104. *The gear-side of a late type 4CD No.34636 (Feb.1925). Originally owned by the Witham Coal & Cartage Co.Ltd. it belonged to J.Shuttlewood of Felstead, Essex, by the time of this picture.*
[R.G.Pratt]

Fig.105. *No.30959 began in 1912 as a tractor with J.Griffiths of Bargoed but ended its working life as a roller with Isaac Ball & Sons Ltd. at Wharles, Lancashire, interestingly converted with non-Garrett parts having a distinctly Aveling appearance. The wider rear rolls, the high (and heavier) fore-carriage head and the larger diameter front rolls were a serious attempt to equalise compressive forces as between front and rear. It is preserved as a showman's tractor.*

[W.P.Riley]

Fig.106. *By contrast the conversion of No.33989 of 1920 was done with Works parts. The tractor/roller spent its life in Ulster with Ballymena U.D.C. When Bob Pratt photographed it he was told the chimney was 'temporary'.*

[R.G.Pratt]

Fig.107. *No.3 in the East Suffolk County Council's fleet, was No.34084. The picture which was taken in August, 1936, illustrates the late type awning and the miniaturised version of the 'Big **G**' applied to rollers. It also brings out the well-groomed appearance it had acquired in the hands of an attentive driver. The painted ring on the flywheel was a shade out of truth, giving a slightly disconcerting effect when in motion.*

[R.G.Pratt]

Fig.108. *No.33425,* The Denmead Champion, *new in Feb.1919, owned by E.W.Edney, Denmead, Hants.*

Fig.109. *Frank Mortlock was noted for invariably wearing a bowler hat - though there is no truth in the suggestion that he wore it to bed. He is posed by his 4CD No.27185 (Aug. 1908) an historic machine which began as a showman's tractor with John Studt and came fourth hand to the Mortlock stable at Lavenham in 1932, only to be sold again in 1934.*

[B.D.Stoyel]

Fig.110. *4CD No.33953 (Sept.1920) in the station yard at Beckermet, Cumberland, in 1946 when owned by W.C.Sutton & Sons. It was newly returned from a general repair at Leiston Works in which it received a new firebox, smokebox, tubes, and front wheel bushes at a total cost of £224. 2. 1.*

[A.J.Martin collection]

Fig.111. *The gear-side of the same tractor at the same place. The tractor still belongs to the Sutton family.*

[J.W.Peirson]

Fig.112. *No.33406, a Ministry of Munitions tractor delivered in Dec. 1918 after the Armistice, seen when owned by Frank Honour of Amersham Common, Bucks, working at tree clearing at New Malden, Surrey, on Jan.11, 1936, on a site now engulfed in urban development.*
[B.D.Stoyel]

Fig.113. *The 4CD had a proven record with heavy hauliers. Here No.32622 (Nov. 1914), when owned by Southern Transport Ltd. of Brighton, is hauling plant to Brighton Corporation's Southwick Power Station.*

Fig.114. *No.33911, new in August, 1920, to K. Miles of Charminster, Dorset, was sold in 1926 to its final owner, A.W. Sheldrick of Toftwood, East Dereham, Norfolk, for whom it is seen threshing on a bitter wartime winter day.*

Figs. 115 -118.
A saga of 1920s out-of-gauge haulage with 4CD No.32740 (May, 1915) owned by Thomas W.Ward. In 1924 the Royal Laundry at Brentwood, Essex, bought a boiler from Wards which was delivered by road. Brook Street Hill on the London side of Brentwood proved more than the tractor could cope with unaided. Accordingly the ex-WD Commer sent to carry the stores and packing was chained to the front (Fig.115) but to no avail. The next attempt was to run the tractor forward, scotch it up and use the wire rope (Fig.116). This failed too as the tractor was pulled back towards the load. The next step was to chain the tractor to the telegraph poles on the London-bound side of the road and rope the trolley up to it. (Fig.117) This worked and was repeated to the top of the hill, thus blocking both carriageways of the A12 whilst it was done. In Fig.118 the outfit stands once again on level ground in the Ongar Road, near the laundry.
[F.D.Simpson]

Fig.119. The saga of Brentwood concludes! The crew have nudged the trolley into position by the boiler house. The sling-off chain between the drawbar and the engine draw plate can just be seen at the feet of the man on the ground as can the timber they have used to give the trolley its final short nudge.
[F.D.Simpson]

Fig.120. No.33990, the North Eastern Railway's second 4CD, (their No.108) supplied in July, 1921, with a Lancashire boiler upon which '30 tons' has been superimposed by the photographer.

Fig.121. Tractor No.28774 posed for a catalogue illustration of its gear-side and drive arrangements, with the pin laid out on a dust-sheet (rt. of pic.). It also shows the inconspicuous position stipulated by von Arx for the Leiston Works plate.

Fig.122. In May, 1934, superheated tractor No.32647 (Feb. 1915) owned by R.R.Harvey of Tannington by then, but formerly with Norfolk Motor Transport Co., under repair at Malcolm Bloomfield's works at Debenham, Suffolk.
[R.G.Pratt]

Fig.123. No.33284 (Oct.30, 1919) had eight owners before David Smith for whom it was threshing at Cox Green, Bucks, on August 8, 1964. [D.R.Smith]

Fig.124. The Oxford Steam Plough Co's. No.33292, an ex-Ministry of Munitions tractor dating from April, 1918.

Fig.125. The load behind the Norfolk Motor Transport Co's. No.32647 (Feb.1915) in this photo dating from 1916 is a pontoon intended for use in the Dardanelles landing.

Fig.126. Another picture from the same year is of Bantam, *No.32680 (March, 1915) owned by Norman Box .*

Fig.127. *Garrett 4CD tractor conversions were popular with showmen during the twenties and thirties. No.33566, new in July, 1919, to Browning Bros. of Stonehouse, Glos. and shown here at Ditchett, Somerset, when owned by W.Cole & Sons of Bristol, was converted by W. Sparrow of Martock, who did much work for showmen.*

Fig.128. *By contrast No.33799 (March, 1920 for Down C.C.) was converted in December, 1924, for Albert Stocks at Leiston Works. The colours were crimson lake, lined black and yellow, and red wheels. It was named* Duke of York. *He had it only until 1929 when motors became more convenient for the work it did.*

Fig.129. *Chas. Heal's* Little Jimmy *(No.33290) was a Ministry of Munitions engine of May, 1918, bought by him and converted in 1920. He used it until c.1936.* [Allan Downs collection]

Chapter 10

Uses and Users

In the individual chapters there have been references to the people and organisations that bought Garrett tractors and rollers, perhaps thereby giving some indication of the magnificent diversity of ownership. The appearance of the rollers preceded that of the tractors by some seven years, and of the total of 386 rollers made only thirty-nine were sold in the British Isles. Half of these were for private operators and the balance for local authorities. Abroad, the position is more difficult to analyse. Seventy-nine were sold through Burn & Co. (Calcutta) Ltd. whose office and works was in Howrah, mostly for road authorities and town councils in North and North-east India, and nearly all were shipped through Calcutta, though No.25558 was sent to Rangoon, whilst No.27724 went to Bombay. As noted in chapter 9, in the post-1919 period the trade to Bombay became mainly the province of John Birch whose London office was at 2 London Wall Buildings, EC 2. In 1920 he had twelve rollers although No.34205 went to Bombay through the agency of Macbeth Taylor & Co. and No.34266 by way of Robert Taylor & Co.Ltd. In 1922 two rollers, Nos.34173and 34174 were shipped to Karachi, the ordering agent in that case being William Menzies. Sometimes the ultimate customer for these Indian rollers is known but for most it is not. Some had long lives. For instance, when the late Steve Neville was working in Calcutta in 1962 he encountered one of them in use by Calcutta Corporation though he could find no Works number on it.

The only Garrett roller shipped to Ceylon, No.32623 for the Colombo Electric Tramways & Lighting Co. Ltd., was ordered through Robert Brooks & Co. whose office was just off Gracechurch Street in the City of London. Two Far Eastern rollers (Nos.32312 and 32610) were sold through the Borneo Co.Ltd. They had been introduced to Garretts by William Marshall, who was mostly concerned with selling CCS and SSW semi-portable power plants, several of which the Borneo Company had. H.F. Roll, the Dutch agent and a friend of Charles Remy, was responsible for orders for a group of rollers shipped to what was then the Dutch East Indies, as well as for a number of examples sold in Holland itself.

However, the Garrett connections with Russia, which then included the Baltic States and much of Poland, long ante-dated Remy's period of influence in Continental sales as did the links with Austro-Hungary but curiously, notwithstanding this long association, only two rollers went to Eastern Europe. The first, 10 ton single No.24400, was sold through Osias Auschnitt of Galatz, Roumania, whose trade was mostly in portables and threshing machines, though he also sold some straw-burning traction engines. Sad to relate, his finances ran into problems and at the end he left the Garretts with a bad debt that had to be written off. The other was No.25374, an 8-tonner supplied to Palmer Graham & Co. of Rostoff-on-Don.

However, it was the German and Swiss-German connection, assiduously fostered by Remy, that sold the largest number of rollers on the Continent, though followed, at not too great a distance, by Italy and Spain. Altogether no less than 95 rollers went to German purchasers. As has been explained elsewhere the system of tariffs operated by Germany to protect its home industry meant that from a tax point of view it was advantageous to import a partly completed roller with substantial parts needing to be manufactured and added in Germany rather than to take in a finished machine. Those who finished off partially completed Garrett rollers to be sold under their own names included Henninger & Co. of Darmstadt; W. & J. Scheid, Limburg a.d. Lahn; Theodor Ohl of the same town; Muscate of Danzig; Ruthemeyer of Soest; and Reifenrath of Niederlahnstein, who later traded as Strassenwalzen-Betrieb.

Th. Pilter, the mainstay of the Garrett trade with France and French North Africa almost from the time he set up in business in Paris in 1864, was another ready-made connection taken, like those in Russia, into the Remy European sales organisation. Though he managed to sell a useful number of tractors and traction engines, there were no roller sales via Pilter. Despite this, as late as 1913 Part One of the Pilter catalogue listed virtually the full range of available Garrett rollers. The failure in selling Garrett rollers in France is curious since Aveling & Porter enjoyed some success there, notwithstanding the presence of successful native manufacturers of whom Albaret was probably the most notable. By contrast, the Belgian agent, E.Weber, whose office was at 12, Square Marguerite, Brussels, is known to have sold at least three - Nos.30301, 30302, and 30313. The other Brussels agent, M. Sigart who had an office near the Midi Station, sold various engines for Garretts but never managed to sell a roller.

The roller trade with Italy was rather more promising and was based upon the firm of Grimaldi, which had been started in Milan in 1864 as Whitmore & Grimaldi. Whitmore, the senior partner, was an expatriate Englishman who died in 1895 before the advent of Garrett rollers. Filippo Grimaldi carried on the business, later handing it on to his son. The Grimaldis, who had expanded their scene of operations to offices in Genoa, Turin, Palermo, and Naples beside the parent office in Milan, found customers for thirty-three rollers. This relative success contrasted with much more limited sales in the Iberian Peninsula. Sumner Harker & Co. (later Whitehead, Sumner Harker & Co.) brokered some sales to Portugal from their offices at 196, Deansgate, Manchester, and in Lisbon and Oporto. The matter of agency in Spain is less clear. The first roller sold to a Spanish customer was No.25028 destined for the Spanish Public Works Ministry and sold through the Sociedad Anglo-Espanola de Motores (the Anglo-Spanish Motor Co.) Anglo-Espanola sold nine rollers in all, their best and final order being for a group of five (Nos.26755 to 26759 inclusive) in December 1907, and January and February, 1908. After that all that can be said with certainty is that a Senor Don Julio Pena y Martin, based at Calle de la Princesa 77, Madrid played a part.

In the case of Spanish speaking America, however, things are more sharply defined. Most of the Garrett trade there was with Argentina where British prestige and investments were both high. The firm had two agents, Agar Cross, 124 Calle Defensa, Buenos Aires, and M.E. Mattos of 299 Calle B. Mitre, Buenos Aires, represented in London by Torromé, Sons & Co. of 25 Lawrence Poultney Lane, only Mattos being involved in vehicle sales. He disposed of twenty-four rollers, for which the Argentinian Ministry of Public Works was the predominant customer. The only other Spanish-speaking country to purchase an appreciable number of Garrett rollers was Cuba, and there the Public Works Department bought five 12 ton compound rollers (Nos.26586 to 26590 inclusive) in July and August, 1908, through the agency of Czarnikow, MacDougall & Co. of 112 Wall Street, New York. Later the firm became the Czarnikow Rionda Company but continued to trade from the same address. Rollers Nos.30907 and 32737 were also sold through this agency. As has been related in the earlier chapters Garrett rollers found their way in limited numbers to South Africa, Japan, Arabia, and New Zealand. None went to Australia, despite the fact that Garrett portables and traction engines enjoyed useful sales there.

Records of the ultimate purchasers of many of the rollers exported do not survive, if, indeed, they ever existed. A few, however, are known by name, of whom the most likely to be recognised by visitors to Dutch rallies is H. Boesewinkel of Zutfen who ordered through H.F. Roll. His first purchase was the 8 ton single No.26556 in 1907, to be followed by two more the next year (Nos.27145 and 27205) and another pair (Nos.27625 and 27934) in 1909. Boesewinkel used these rollers for many years and some of his rollers, though none of the Garretts, survive in preservation.

In most cases, however, exported rollers effectively disappeared from view once they were shipped. At wide intervals an order for spares reached Leiston Works but otherwise nothing was known about either the machines or their owners. The position with tractors, however, was quite different. Instead of 90% going abroad, 85% remained on home territory, either new or second-hand, in the hands of an extremely diverse group of owners. In the latter years of steam the Garrett tractors were mostly engaged in fairground use, agriculture, or in round timber haulage, which often included its extraction from pasture land or woods, but there were some other important or interesting aspects, notably the long survival of the group of steam tractors in the fleet of John Harkness & Co. in Belfast. That the fleet should have lasted in very heavy service until the mid-1960s was the result of a series of factors too complex to be treated in detail here. In one sense the curious economic situation in the six counties and the concentration of most of the available heavy haulage work into the city and environs of Belfast made anything larger or faster unnecessary whilst the absence of effective competition discouraged the adoption of more modern plant. The willingness of Leiston Works to continue the supply of spare parts also made a difference. During one of my visits to the Works when writing *Garretts of Leiston* my late friend Ted Dunn was telephoned by Charles Harkness about the availability of a particular part. "How long are you going on running the tractors?" Ted asked. "As long as you can keep us supplied with parts" was the reply.

The tractors latterly in service were Nos.33414, 33419, and 33545, all of which had been converted to pressed-on solid rubber tyres on replacement wheels supplied by Garretts. These ran until late 1965, but over the years Harkness had had a total of twelve 4CD tractors plus the 6NHP road loco *Vera*. Apart from their rugged ability to stand up to knocks and the punishment of sett roads and yards the quality of the tractors that stood them in best stead was the capability of carrying 75 yards of $^5/_8$" diameter rope on the winding drum. It was these characteristics that also appealed to round-timber hauliers. Whilst it can be argued that most of the tractor builders would have been prepared to make a tractor with a larger than standard drum had they received a special order timber hauliers, on the whole, were an impecunious group seldom buying their plant new but rather acquiring what other more affluent trades had had the first use of. Of course there were exceptions to this statement. For instance, Samuel Rea of Antrim, who had tractor No.28391 (and, subsequently, Nos.33086 and 33633 also), owned a large sawmill, later powered by a Garrett CCS power plant - all these engines were bought new. He was also a coal merchant who, like many merchants at that time, was willing to supply coal from which the smalls and dust had been screened, the CCS providing a convenient and economical way of using the screenings.

At least two other coal merchants trading on a large scale became owners of Garrett tractors. Thomas Moy Ltd., whose headquarters were in Colchester but who traded over much of Essex and Suffolk, had superheated tractor No.31248 in 1913, using it from the Ipswich branch for some years before parting with it in favour of steam wagons or motor lorries. E. & A. Shadrack of East Ham, however, had No.32653 in February, 1915, for only about a year before selling it to John Mowlem & Co. Ltd., the Westminster building contractors. Like Thomas Moy the Shadracks found wagons more convenient. No.3 tractor No.25399 was tried by the Carr Brothers who were coal merchants in Leiston and also owned the brickworks. Talbot Carr rejoiced, if that is an appropriate term, in the nick-name of Rungie. Despite being on friendly terms with the Garretts he and his brother decided to continue with horses for haulage.

To return from this digression to the theme of heavy haulage whilst it is fair to say that no other user attained such a record for continuity of use or originality of methods as Harkness there were others involved in the heavier type of work. A tractor thus used was No.32740 owned by Thos. W. Ward Ltd. and based at their Silvertown (London) depot. This was one of the few turned out on Bauly Tangent wooden wheels and, moreover, had wood-block treads on the rear wheels, though straightforward steel tyres at the front. This tractor and its crew were given the task of delivering a new 25 ton boiler from Silvertown to Brentwood, Essex, and positioning it in the boiler house of the Royal Laundry in the Ongar Road. This was owned by the father of my friend Frank D. Simpson, and a sequence of photographs exists showing the events that took place at the end of the journey (Figs.115 to 119). The journey went slowly but without untoward incident until the foot of Brook Street Hill on the A12 just south of Brentwood was reached. Here the tractor jibbed. An ex-War Department Commer lorry had been sent with the outfit to carry the jacks, blocks, packing, coal, and other impedimenta. This was coupled up to the front of the tractor and the double headed procession moved a little way up the hill only to halt again. This time it was decided to scotch up the trolley, drive the tractor forward, run out the wire rope and winch the load up to the tractor. The reverse occurred. The tractor was undefeated, but since the resistance of the load was greater than the tractor's adhesion on the road surface the tractor began to wind itself back to the load rather than vice versa. However, nothing defeats a heavy haulage gang. The telephone lines were carried up the east side of the road on pairs of heavy poles. Notwithstanding the fact that they were on the A12 and that even in the mid-twenties the traffic on it was not negligible it was decided to place the tractor at an angle across the London-bound side of the carriageway, chain the front of it to a pair of poles and with it thus anchored wind up the trolley to it. As it neared the tractor the trolley was more or less astride the centre of the road, traffic being held up in both directions. Nevertheless it was scotched up whilst the tractor ran on to the next pair of telephone poles where the process was repeated and so on to the top of the hill. Frank commented that Ward's crew were totally unperturbed by the disturbance they were causing to the traffic, and the police found it convenient to be elsewhere. Apart from No.32740 Ward's Silvertown depot had three other Garrett tractors, the earliest of which was No.32084 (March, 1914). Like No.32740 it was mounted on Bauly Tangent wheels with wood blocks to the rear rims and plain steel tyres at the front. This tractor was pressed to 200 p.s.i. In 1924 they bought No.34410, their third tractor, forsaking the Bauly wheels for Garrett wheels with pressed on rubbers all round. This tractor was ordered with 'show finish throughout' in a livery of crimson lake with gold lining. The last tractor bought by Wards, also on rubbers to all wheels and finished in the same livery as No.34410, was No.34787 which was one of the Garrett exhibits at the 1926 and 1927 Norfolk Shows and had thus been in stock for some while before being sent to Thomas W. Ward in July, 1927.

The pattern of No.4 tractor sales was entirely different to that of steam rollers. Whilst it is true that most of the No.4 singles went abroad they were too few in number to be of significance. In the case of the 4CDs only just under 12% of a total of 515 left the British Isles. By far the largest order for tractors was that from the Ministry of Munitions in 1918 but these hardly stirred in military service and within a year or two were added to the numbers already in non-government ownership. This apart, the customers were predominantly in the private sector, few tractors going to local or central government.

It tends to be overlooked how short the hey-day of the steam tractor was, and particularly that of the Garrett output which did not get into annual double figures until 1906, reached a peak in the three years 1918 to 1920, and then fell away more or less to nothing. By about 1923/24 the position of steam tractors generally was already being actively eroded, on the one hand by developments in steam wagons and on the other by the inroads of the internal combustion engine. In other words tractor building and usage peaked for not much more than a decade. By the time of Sir Arthur Salter's now notorious report upon the taxation of steam road vehicles and the prohibitory rise in licence costs that flowed from it, the steam tractor was already in eclipse. Typically by the time the 1930s arrived the widest spread of tractor use was in the three trades of the fairground, threshing, and round timber haulage. A contracting market made the prices of second-hand tractors economically accessible to prospective purchasers in all three categories. This state of affairs applied, of course, to all makes of steam tractors and not merely to Garretts although some proved to be of more enduring popularity than others.

The late Syd Sutton of Beckermet, in Cumberland, worked in and eventually took over his father's business that embraced threshing, a certain amount of round timber haulage, tar spraying and rolling. In 1927 Syd's father bought No.33953, new in 1920, to Long Ashton Rural District Council in Somerset. The haulage part was the first to drop out but he found the Garrett 4CD ideal for the remaining functions, particularly threshing. He also had No.33636 which was converted to a roller in 1950. Many of his threshing clients had farms with very restricted stackyards, and a high proportion of these were approached over the narrow and steep lanes of the Lake District. The compact proportions of the No.4 tractor made it very handy for this sort of use. Syd once described this work very graphically when he wrote:

> I must say that threshing in the Lake District was no easy job: it was not uncommon to have to set up in three stackyards in one day, and occasionally move into the fourth for next morning. Most yards were awkward and narrow, as were the approach roads to the farms, and sometimes I've had to take off the wheel caps of the machine to pass between dry stone walls, because if a cap touched one cobble most likely a few yards of wall would collapse.
>
> We used various engines for threshing, but I preferred the Garrett No.4. She was a great little engine in every way, for with tight rings, accurate valve settings, a clean exhaust pipe, and above all a driver who had his heart in steam, she performed beautifully, and now when I look back I would not have exchanged my job with anyone. Incidentally, there were always three men with a set of threshing tackle, which comprised engine, thresher, straw trusser (fitted to machine with raising and lowering gear) and a living van, whilst in later years we also took a stationary type baler.
>
> Over most of the years 33953 drove a Garrett thresher No.15990, new in 1929 (this was the only wholly Garrett set in Cumberland). The machine was heavy but she handled it perfectly. Occasionally I had to turn the engine and push with the rear in some of the difficult yards, and also in later years when we took a stationary-type baler we had to split the load up on one or two of the very steep hills. On one or two occasions I had to set the engine backwards to the machine because the gradient of the yard was so steep. This was quite simple; one of my mates would stand on the front tank and look along the face of the flywheel and tell me what was required. Unfortunately I never got a photo whilst set up like this; in fact I have very few shots taken when threshing. I suppose like many more people I expected steam threshing to go on and on.

The family business using Garrett tractors with which I was most acquainted was that of Tom Newell at Westerham, Kent, whose sons Jack and Bill also worked in the business, though not, to the best of my knowledge, as partners. At various times Tom owned three Garrett tractors, though not simultaneously. When he first set up in business on his own account in the aftermath of the 1914-18 war he bought a Wallis & Steevens tractor (No.7214) but this having failed to come up to his expectations he sold it in 1924 to Chris Lambert at Horsmonden, and purchased instead Garrett No.33078 (1918), one of the tractors bought new by the Metropolitan Asylums Board. I remember him relating that when it arrived he found to his chagrin that it had been daubed with red, white, and blue paint to take part in a hospital fete. Its first journey, therefore, was to Thomas Wood & Sons Ltd. at nearby Crockenhill for a repaint. Three years later he bought No.33422, also built in 1918, from the estate of Archibald Maconochie, together with an Aveling & Porter GND tractor (No.7640). He divided the work so that the Garretts worked, as far as possible, on snigging out and loading the trees, at which their long wire ropes and directly driven drums made them particularly useful, leaving road work to the Aveling, which on the whole he found faster. By 1934 the Aveling, which had been built in 1912 and had endured seven years hammering by Tom, was sold, but he bought two more (Nos.6831 and 7880) and another Garrett, No.30050, which was new to Noah Etheridge of Blythburgh, a few miles from Leiston, in 1911. This was pretty well worn when Tom bought it, and I suspect that it did little work but was robbed for parts. Tom Newell put up a dogged battle to keep going entirely by steam against the competition of rivals who had gone over to diesels, but in the end had to give in and purchase a Foden diesel. This meant the end for the Avelings but at least one Garrett, I believe No.33422, was kept going until the late forties.

I have no 'play-by-play' description of what Tom's Garretts had to suffer but I do have the account of a Norfolk driver (Mr. J. Aldrich) of an episode at the end of the first World War when he was driving No.33067 for the Norfolk Motor Transport Co. of Norwich:

> I had three elms on a drug sixty-three feet long, from Woodton Old Hall; they weighed eleven tons, and were loaded with a Clayton crane engine. After they were loaded, I asked Jack (the driver of the crane engine) to help me pull them out to the road, as I only had a Garrett No.4 tractor and I had to come out between two ponds, but he wouldn't. So I got plenty of steam, and she pulled them all right until I got to the ponds, where it was a bit uphill; so I took the tractor off, and I had two 15ft. chains with which I chained the front of the tractor to a tree on the bank, and blocked the back. When I pulled the rope broke, so I tied a knot in it but that didn't hold, so I had to splice it. I just don't know how she shifted it but she did, and it was eight o'clock at night before I got on to the road; so I decided to leave the drug at the 'Dick' public house, and stop in the van for the night. When I pulled in the pub yard, I swept the hedge down!
>
> The next morning I got on the road for Norwich City station, and they were steam rolling just through Woodton and the tractor scrabbled herself in on the road. I asked the driver of the steam roller to help me out, but he said he was afraid he would pull the tank off. So I ran the tractor forward, blocked her and put a chain from the pin to the back wheel, and roped the load again, which brought the comment from the roller driver, 'You've got a good bit of stuff there.' As I turned into the station yard at Norwich, I was told by my partner that the trees only missed the windows by inches.

Whilst this particular incident may have represented the extreme of what a timber tractor had to endure the life in general was rough.

One of the Newell Garretts driven by Jack Newell was working on extracting timber from a hilltop wood at Manor Farm, Sundridge, Kent, involving weaving the loaded tug down through an orchard to a sharp left-hand bend at the bottom of the slope where the track continued along the bank of a brook. Habitually a log was trailed on a chain behind the tug to give retardation. On the occasion in question this log hit a stump, breaking the chain. The load pushed the tractor downhill, gathering speed. By dexterous steering Jack kept it clear of the fruit trees but at the bottom he had a stark choice. Either he could go straight on into the stream with the tug on top of him or attempt the left turn and risk being pushed over. Jack chose the latter, the tractor was rolled over, as he more or less expected, and he was thrown clean through the roof. He suffered no broken bones and the tractor had only a broken chimney and a slightly buckled rear wheel. After being pulled upright by the Aveling it went back to work.

By the time Garretts set up their system of travelling inspectors, of which the veteran Tom Staulkey became the first, the output of steam tractors had become very small. In respect of tractors, therefore, we do not have anything like the number of illuminating and amusing reports that he made upon the relatively new undertype wagons and their owners which he visited. One of the few upon visits to tractors related to No.34702 delivered to Coltman Brothers, the timber hauliers of Claybrooke Magna, near Rugby, in March, 1925. This carried 75 yards of 5/8" rope and was fitted with 14" wide hind wheels. The Coltmans had it painted in maroon and lettered in gold leaf. The Coltman firm was no makeshift outfit but the owners expected a high output for the money they laid out. Tom called to see the tractor in May, 1925, when it was only some two months old. All seemed to be well as his report made clear:

> The above are well pleased with their tractor in every way. It is working every day and in very bad places to get their timber out.
>
> I went out with one of the above, and met the engine on the road with a load of timber. The engine is running alright.
>
> The injector they say does not work clean at 200 lbs., but I found that the water was very hot in the tank. This was caused by lifting water and also the driver having the plug cock attached to waterlifter too slack in the plug. I adjusted this for him. Otherwise the driver seems a capable man, as he drives and steers alone.

Things were not quite so bright by September, however, as the tubes had begun to give trouble by persistent leaking at the firebox end. Because Tom could not, or would not, drive anything not propelled by steam his movements were all made by public transport or in the cars of clients or Garrett representatives. On September 18, therefore, he was taken to Claybrooke by H.R. (Jack) Simpson, by that time the Garrett representative for the Birmingham area.

Mr. Coltman tells us he is having trouble with the boiler tubes. This I am not surprised at, as his driver does all the work of stoking and steering. This means that at times he has an enormous fire, and before he stokes again cold air would be passing through. I pointed this out to Mr. Coltman when I visited him last May. He quite agrees this may be the case, and I have suggested that he persuades his driver to teach his mate the steering, so he can better attend to the boiler.

The high pressure connecting rod brass has also given trouble, this apparently has been hot and could not be properly rebedded. Unfortunately the engine was working some distance away, so I could not see it.

Mr. Coltman thought he should have his tubes ferruled, but I have suggested to him that he have a boilermaker to attend to these, and with my suggestion as regard the driver, will experience no trouble.

The tribulations that resulted from this sound advice were detailed in a further report from Tom dated January 26, 1926:

I saw the above customer yesterday the 25th and they were greatly upset with the trouble they have had with the boiler tubes of the above tractor. As the tubes had considerably furred they decided to have them drawn and cleaned, and engaged a boilermaker (Mr. Westony, of Rugby) to do the job. After working some time at the job he found that the tubes would not draw out owing to them being too deeply expanded at the smokebox end, and gave the job up, telling Messrs. Coltman that Garrett's ought to do the job. They then acquainted Mr. Simpson, and arranged for Midland Rolling & Haulage Co. to undertake the job. Their boilermaker arriving he found the same trouble, and had to get an Acetylene plant, and burn the tubes from the inside, driving the tubes into the boiler. Mr. Coltman tells me the tubes have never given trouble at the smokebox end. Seeing the driver some time after I also questioned him, but with the same answer. I cannot think these tubes were expanded so deeply at the Works, and impossible if the expander is used with the shield. I have 2 pieces of tube which I will send to the Works, one plain end showing the raised part, and one screw tube end, which they say was broken, and holding on to only a small part when tube was burnt through.

I saw Mr. Simpson this morning, and suggested to him that I visit the Midland Rolling & Haulage Co., which I have done. Unfortunately their boiler maker was out, and their foreman, Mr. Smith, could not give me any further details, though of course he knew the difficult job they had.

I thought that probably as the driver was working with him, he might have known under what condition the tubes had been expanded.

The problem with the Coltmans' tractor, as with many others in the same trade, was that it was engaged in a rough and demanding business, offering hard work, poor conditions and little personal ease. Men who could tolerate the conditions imposed could not be expected to be over considerate of the machinery they worked with. The tractor had only a short life with Coltmans before being replaced by a new Foden steam timber tractor. After a year or two, in its turn, this came to be replaced by a diesel and then passed into the hands of Esmond Kimbell who adapted it for threshing. The condition in which this relatively new engine was when they got it caused a good deal of tut-tutting by George Wright, the Kimbell fitter.

A complaint about tractor No.32946 was dealt with not by Tom Staulkey but by J.K. Peecock who had charge of the Test house in AGE days, a man of profound knowledge and infinite resource. The tractor was supplied in October, 1916, to Mr. E.C. Keith of Castle Acre, Norfolk, (later of Swanton Morley) who had put it to work driving a 54" threshing drum. After a lapse of nine years a complaint reached the Works that the tractor would not drive the drum fast enough, and never had done so. Why the owner waited nine years in a state of festering discontent before making his dissatisfaction known one will never know. Possibly a Garrett salesman tried to sell him some other item and received a reply he had not expected, but whatever the source of the information Peecock, renowned as a trouble-shooter, was asked to investigate. His report on this visit and its outcome was dated October 1, 1925, and read as follows:

The complaint made was that the tractor would not drive a thresher fast enough, and never had done. This was not to be expected as, on examination, the governor valve which, normally, should be screwed tightly into the cylinder, was so loose that I could turn it round with my thumb and finger and screw it right down into the steam chest without tightening it.

It was therefore necessary to fit another valve. This I did, and on putting the new valve on the governor spindle I found the boss through which the spindle passes to be 3/16" thicker than the original one. This, of course, had to be turned down to the original thickness before it could be used, as the split pin hole in the spindle was completely covered by the nut, which was only 1/4" thick.

When this was done and the tractor started threshing, it was found that it threshed equally as well as a 6 H.P. S/C Ransome traction which had been hired whilst the Garrett tractor was out of commission. They were very surprised, as they told me previously that they were sure the engine was not powerful enough for the job. They also said that the tractor had never made the thresher 'hum', but it 'hummed' before I left it and they heard it over 200 yards away.

To return to Tom Staulkey's travels it is perhaps worth mentioning, by way of contrast to the dissatisfaction of Mr. Keith over No.32946, the very evident satisfaction felt by R. & A.J. Rose of North Luffenham, Rutland, the owners of No.34713. The Roses obviously expected a good performance as a matter of course. Tom Staulkey noted:

While at Luffenham I had the opportunity of seeing the above customers, and tractor in the Station yard. They speak very highly of the tractor and are doing plenty of timber carting. The only complaint they have is that the two front guide rollers have large blow holes in them, and the wire rope catching these holes prevents the roller from turning. They are too large to be stopped up satisfactorily.

The only convertible he is known to have visited was No.32923 which had been supplied to Cornwall County Council in October, 1916. What he wrote about the visit on October 13, 1925, intrigues me, mainly because I have not been able to fully see the reason for the problems of which he made mention.

The above is being used now as a road roller. The boiler was being washed out when I was there, and the driver informed me that the engine works quite satisfactorily. The driver mentioned the trouble he had with the firebars bending in the long firebox as they have the firebars made in the district, they have 3 distance blocks along the bar instead of one in the centre.

The driver was telling me he could not use the brakes when rolling in the inside rollers, as they are not true and the frost peg bosses project too far.

Clearly Cornwall County Council, like many other owners of convertibles, had decided that frequent conversions were far too time consuming.

Visits to rollers were not common either. On July 8, 1925, Tom went to see Zachariah Fairclough at Clacton, who had had the 10 ton *Baron* roller No.34706 on March 31 that year. His expectation of seeing the roller was confounded but he did have the pleasure of meeting a highly satisfied client. One would love to know more about the apparent minor problems referred to but the opportunity is now long gone.

I visited Mr. Fairclough on July 8th, and in conversation he speaks very highly of his road rollers, and especially of the economy in coal with his compound rollers as compared with his 10 ton S/C which he has had many years. The last new compound roller is working for the E.S.C.C. not far from Ipswich, and as regards the main axle not getting sufficient oil, there cannot be anything wrong as the roller has been working ever since delivered and Mr. Fairclough has received no complaints.

The oil caps on gudgeon pin lubricators are not now supplied.

As regards the flywheel going on to the crank further, this no doubt is put right by the driver.

Mr. Fairclough hoped if I was anywhere near Ipswich I would see his roller. He was very pleased at my visit.

A Garrett man through and through, Tom Staulkey never lost the chance of passing on a hint of a potential sale or of paving the way for one, as witness his visit to R. Dingle & Sons at Stoke Climsland on October 7, 1925, Mr. George Dingle, to whom he refers, had taken the lead in the negotiations that had led to Dingles being granted the status of Garrett agents, and it was he also who had discussed with Schmach and his draughtsmen the refinements he had required in their successive purchases of rollers.

I was at Stoke Climsland yesterday the 7th inst. Mr George Dingle, the gentleman who was often at Leiston, is away on holiday, his brother Edmund John Dingle was also gone to Callington. On returning to that town I met Mr. E.J. Dingle and had a very interesting conversation with him. He speaks very highly of the Garrett rollers and tractors.

I was introduced to Mr. J.H. Dingle, Contractors and Builders, Kelly Bray, Callington, who have a Garrett tractor, purchased from R. Dingle & Sons after the war. It is an old type tractor, with fine tooth spur ring and pinion, old type belly tank etc. There was no number plate on as it had been removed to fit new type starting gear bracket. This engine is in good working order, and they are very pleased with it. Mr. J.H. Dingle is an enterprising young man, and speaking of tractors on rubber tyres, he says he could do with rubber tyres 9 months out of 12. I think he would seriously consider a quotation for a set of wheels.

George Dingle, whom Tom Staulkey just missed, was a most forthright man who took a great interest in roller design. After he had had No.30074, equipped with the new design of swivel fork, he wrote a characteristic letter, dated February 22, 1912, to Frank Garrett, Senior, voicing his misgivings:

No doubt you will remember when I was at Leiston that I spoke to you about the front fork arrangement of the 10 ton roller, that I considered there was a weakness where it was hinged. There being a shortage of bearing surface to resist the leverage of the front rolls. I visited the roller at work yesterday and I found that it is nearly $^1/_{16}$" slack with the pin in the fork (not the central spindle where it is supposed to rotate). We have already been in correspondence with the firm respecting a flaw in the fork but if another fork is sent to replace it, it will not be satisfactory on account of the reason mentioned above. We should far rather prefer the old pattern cast iron fork and saddle with all its shortcomings (which I mentioned to you) than the present arrangements as we are certain it will be an endless trouble. The bearing surface of the hinge pin ought to be at least 20" long to resist the leverage and length of the centre spindle that fits in the saddle ought to be increased. We should be obliged if you would give the matter your consideration and as already mentioned we would rather have the old pattern fork and saddle with all its failures than have the present thing unless it is considerably altered.

His complaints were listened to and some small dimensional adjustments were made to the swivel fork, but the new design of fork met with general approval especially with overseas customers and remained in use to the end without any of the catastrophic consequences predicted in the letter.

By far the greater number of Garrett tractors were bought new by various categories of haulage contractors. Some, such as Harkness, Coupe Bros., Southern Transport, or Norman E. Box, engaged in heavy haulage; others of the type of Tom Everton, were general hauliers; and others, again, like Lancaster, Robert Wynn & Sons, or Coltman Brothers undertook round timber cartage. None of these could be described as national names with the exception of Norman Box whose assiduity and dexterity in promoting the image of his firm made him known more widely than was merited even by his audacious and meticulously planned movements of large indivisible loads. These rarely involved his Garrett tractors except possibly in hauling the living van or the stores truck. One firm that might be accorded the status of at least a semi-national name was Sutton & Co. of Guildford, Surrey, who purchased Nos.33025 and 33221 and were parcel carriers over a considerable area of the Home Counties and beyond.

Unlike the overtype wagons few tractors found owners in the brewing trade. So far as I know, the only one to have been owned by a brewer was the No.4 single (No.26515) which went to W. & T. Jenkins Ltd. who owned Brynsadler Brewery in Glamorganshire, but the four tractors, (Nos.31751, 32357, 32647, and 33067) owned by the Norfolk Motor Transport Co. of Norwich, had tentative connections with brewing. This firm did contract carting of bottled beer for Whitbreads, mostly using their steam wagons, but doubtless, from time to time, the tractors became involved in the work. A few tractors were employed in mineral extraction and treatment. Nos.31926, 32559, and 33342 were owned by the Cornish Tin Smelting Co. Ltd. of Redruth. Two were owned by coal mining companies, No.32730 by the Fife Coal Co. Ltd. of Leven, Fife, who had three 5/6 ton trailer wagons with it, and No.33608 by the Thomas Merthyr Colliery Ltd. of Waunwyllt Colliery, South Wales. Perhaps it is irrational but one would have expected colliery companies to have had more. One of the more unusual purchases was that of tractor No.33636, later one of the Sutton stable at Beckermet where it was converted to a roller. This was bought new by the Ecclesiastical Commissioners, whose office was at Millbank, London, S.W.1 and who had it consigned to Canterbury West Station but for what purpose is not recorded.

At least three 4CDs are known to have been used in the furniture removal trade. In October, 1908, Freston Brothers of Ipswich had No.27164, only the twelfth to be built. It seems not to have been a success with them as by 1910 it was in the hands of R. Layton Hitchcock of Bury St. Edmunds. The second to be so used was No.31110 supplied to John James of Merthyr Tydfil, Glamorgan, via the agency of Frank Munn. The third was No.33828 which was sold to John Mellor & Sons of Guide Bridge, Manchester, in July, 1920, equipped with solid rubber tyres on all wheels. I have no other recorded owner of it until 1932 when it was owned by Tudor Jones in Denbighshire, giving it a life of some twelve years with Mellor & Sons, not a bad score when one considers the advances the motor lorry had made into furniture removal work during that period. Doubtless numerous other 4CDs moved furniture at some point during their working lives but I know of none beyond these three that did it full time.

Of the 515 4CDs made 59 were exported and 82 went to either the Ministry of Munitions or the War Department. As might have been expected East Anglia and the Fens (i.e. Essex, Suffolk, Norfolk, Cambridgeshire, Soke of Peterborough, and Fenland Lincolnshire) accounted for the highest proportion of first owners - 102 in all, followed by Ireland with 67, 22 of which were sold in Belfast. South Western and Western counties (Hampshire, Wiltshire, Dorset, Devon, and Cornwall) had 35, Wales had 29, but Scotland only 17. Greater London purchasers had 15 whilst Lancashire, Yorkshire, and Cheshire between them took 19. On the other hand what may loosely be described as the South Midlands and the Welsh Border counties - Oxfordshire, Gloucestershire, Staffordshire, Worcestershire, Warwickshire, Herefordshire, and Shropshire - had 39. In contrast the Home Counties - Buckinghamshire, Bedfordshire, Hertfordshire, Surrey, Sussex, and Kent - had only 16 between them. If this distribution could be described as thinly scattered that in the remainder of the country was even more so. By the later standards of the motor industry all the figures are, of course, pathetically small. Even so the firm did enjoy loyal support from customers. Harkness, Box, and Ward have already been mentioned, but there were others, often in a relatively small way of business, who were faithful to the 4CD. Benjamin Hillen of Woodham Mortimer, Maldon, Essex, for instance, had four tractors between 1913 and 1920; Noah Etheridge of Blythburgh, Suffolk, had five, and W.S. Dewing of Norwich had four. In Belfast Curry & Dodd had seven, almost rivalling the Harkness collection.

None of these instances arose from large single orders, such as those placed by Mattos for the Ministerio Obras Publicas in Buenos Aires, but were deliberate repeat orders after sufficient time had elapsed for the tractors first acquired to have been proved in service. As such they constitute the best possible testimony that the 4CD was an excellently built and usefully proportioned machine capable of employment in all kinds of situations varying from a rural threshing round to heavy haulage in major cities.

__Fig.130.__ It is probable that this standard 10 ton 4-shaft single was No.25054 (new Oct. 1904) for Mysore Corporation via Burns of Calcutta. This was fitted with a Bomford scarifier at the Works. The negative number 1004 (in the earlier series) agrees with the date of No.25054.

__Figs. 131 and 132.__ Playle's tractor No.34021 (1922) was used for a while as Works transport before being sold as 'second-hand', and its crew demonstrates the tricky art of 'rolling 'em up' in which sizing up the weight of the logs and placing the chains accordingly is critical to success. Round timber haulage and the fairground (or the two combined as with the Harris family of Sussex) were occupations that kept a good many steam tractors going when other uses ceased.

Chapter 11

Characteristics and Preferences

A reading of the preceding chapters may well have made it clear that whilst at any one time there was a basic or standard design of the essential parts of a tractor there were virtually no limits to what would be done by way of varying details to meet a customer's requirements. Colour and lettering, of course, were the two items most commonly changed in this way, and a few extra decorative flourishes executed gratis provided a convenient means of giving a prospective client the feeling that he had been made the beneficiary of a very personal discount.

I have explained in other places how the workings of the Agricultural Engineers Association (AEA), formed to combat the effects of the agricultural slump in the late 1870s, and of which all the major traction engine builders were members, acted as a price fixing mechanism on home sales of traction and portable engines and other items of agricultural machinery such as threshers. In the early days of tractors their absence from the price schedules of the AEA gave manufacturers such as Garretts the chance to put a competitive edge on their prices for tractors but this omission was soon amended. Since makers wished the AEA scheme to curb their competitors rather than themselves, ways of circumventing the ban on discounts were eagerly sought, making most firms grudging observers of the letter of the agreement rather than participators in the spirit of it. I do not have a note of when tractors were first brought within the scope of the agreement but believe it was c.1910/11. The table below shows how this pricing system operated as from December 1, 1923, relating the prices of tractors to the scheme long applied to tractions.

STEAM TRACTORS

	5 Ton	7¼Ton
Compound Steam Motor Tractor with winding forward drum, 50 yards steel wire rope, steam water lifter and 26 ft. suction hose, pump and injector or two injectors, front tank, side covers for motion work, differential gear, special brake acting on both driving wheels, driver's seat and outfit, extra large coal bunker, short range spring gear to front and back axles, short awning.	**£**	**£**
(Complying with the existing 1923 Regulations) **Price**	**800**	**940**
Excepting present 5 ton types of **Foster, Fowler, Ruston , Ransomes and Wallis,** the price of which is £750. These to be described as **"Light Type"**.		
Single Cylinder Steam Motor Tractor (complying with existing 1923 Regns.) **Price**	**705**	**845**
Awning over tender may be supplied without extra charge, also sling chain or hauling chain, and bottle jack.		
Extras:- Full Length Awning	5	5
Governors	17	17
Superheater	83	83
Tubular Waterheater	9	9
Brake on Flywheel	9	9
Third Speed Road Gear	42	42
Rubber Tyres for Tractors (in lieu of ordinary type)		
For 160 mm. *i.e.* £98 for Hind and £28 for Front Wheels	126	126
For 140 mm. *i.e.* £88 for Hind and £17 for Front Wheels	105	105
For 120 mm. *i.e.* £70 for Hind and £14 for Front wheels	84	84

The term 'outfit' was defined as follows:

The following is the maximum equipment which may be supplied with any Agricultural Traction Engine, Tractor, Road Locomotive or Road Roller. Signatories may work up to this at their discretion, but any article of outfit supplied other than those in this list to be charged for at a proper market Price:

1 gallon oil tin.
1 ,, engine oil.
½ ,, oil tin.
½ ,, cylinder oil.
1 oilcan.
3 lbs. tallow.
1 tallow kettle.
1 set of firing tools.
1 clinker shovel.
1 mud rake.
1 tube brush and rod.
1 set of spanners.
1 moving key.
1 bucket.
1 funnel.
1 hand hammer.
1 crowbar.
2 cold chisels.
1 caulking tool.
1 file and handle.
2 lbs. assorted gland packing.
4 lbs. cleaning waste.
6 sheets emery cloth.
2 lbs. red lead.
3 gauge glasses and 6 rings.
1 fusible plug.
3 doz. assorted split pins.
3 lamps (2 head and 1 tail.)
1 waterproof cover.
1 set of 16 spuds with pins and cotters (for wheels of 6 ft. 6 ins. diameter or over, 24 spuds with pins and cotters).
1 set of 32 frost studs.
1 drag bar and pin.
1 haulage pin and cotter.
1 water lifter, hose pipe 26 ft., and one strap for do. (except for road rollers).
1 sling chain or hauling chain.
1 bottle jack.
1 governor driving belt.
1 toolbox with padlock and key.
1 instruction book.
1 synopsis of Traction Engine Law.

In a third schedule these prices were related to those for convertibles:

		3 N.H.P.	4 N.H.P.
Actual weight of Roller when converted, without coal and water	tons	8	10
Price of Single Cylinder Traction Engine, including Winding Forward Drum and 50 yards of Steel Wire Rope, Differential Gear, Steam Waterlifter and 26 feet of Suction Hose and Outfit (Governors not included)	£	549	689
Extras constituting Tractor—			
Short Range Spring Gear (hind and front) ..	£	82	82
Front Tank	£	29	29
Short Awning	£	21	21
Enlarged Bunker	£	9	9
Rim Brakes	£	15	15
Single Cylinder Tractor ..	£	705	845
Roller Parts for same, including Hind Rollers, Front Rollers, Roller Forecarriage Bracket, Forecarriage Fork and Axle, Scrapers and Gear, Steerage Chains, Side Covers for motion work, Solid Flywheel or Cover for Flywheel, Injector and connections	£	213	213
	£	918	1058
		TRAC	TORS
Price of Compound Cylinder Traction Engine, with fittings as above	£	644	784
Extras constituting Tractor—			
Short Range Spring Gear (hind and front) ..	£	82	82
Front Tank..	£	29	29
Short Awning	£	21	21
Enlarged Bunker	£	9	9
Rim Brakes	£	15	15
Compound Tractor ..	£	800	940
Price of Roller Parts, same as above	£	213	213
	£	1013	1153

Extras:

Spark Arrestor on Chimney Top	£6	Spark Arrestor (plain wired top)	£2
Awning over Engine and Tender	£26		
Coal Rack	£9		
Flywheel Brake	£9		
Copper Pipes	£9		
Long Range Spring Gear, hind end	£99		
Third Speed Gear	£42		
Governors	£17		
Cast Steel Rims for Hind Rollers	£65 (3NHP) £68 (4NHP)		
Cast Steel Rims for Front Rollers	£59		
Cast Steel Rims for Hind and Front Rollers	£124 (3NHP) £127 (4NHP)		

Long Range Spring Gear applies to any arrangement in which the gear centres and pitch lines are rigidly fixed.
Short Range Spring Gear applies to an arrangement in which the centres are movable allowing an 'in and out' movement of the teeth.

When the schedules were next revised in December 1926, the price of a 5 ton tractor had been reduced to £750 and of a 7¼ ton to £879. It seems not to have dawned upon those prescribing these minimum charges that the years of ringed prices for tractors had come to an end and that the time had arrived when the market place and not the ring had control of prices. When No.34787 was sold to Thomas W. Ward in July, 1927, on pressed-on rubbers all round, the price charged should have been £992, i.e. £879 for the tractor plus £113 for the additional cost of the tyres. In fact it changed hands for £787-10-0. In effect tractors had become unsaleable and manufacturers with stocks were glad to clear them at what price they would fetch.

In the opening period of their tractor building Garretts did not fit an awning as standard. Tractor No.26479, the first 4CD, was quoted for and delivered without a canopy, but thereafter a short awning seems to have been included in most tenders. On the other hand the first No.4 single, No.25646, *was* fitted with an awning, derived from those designed for the rollers up to that time. The earlier 10 and 12 ton roller awnings were substantial, albeit not decorative, structures framed up in selected pitch pine and covered with ½" thick square edged boards topped with 24 gauge galvanized iron sheeting. The overall dimensions of such an awning were 9' 0" x 4' 11", a width that might well have given protection from the sun in India, where most were sent, but not much against rain. The main cross members to which the flattened ends of the supporting 1½" diameter iron supporting standards were bolted were cut from 1½" x 6" material. The four intermediate ribs were 1" thick but cut on the curve top and bottom so as to finish 2¾" deep. The roof boarding overhung the fascias which were 1½" x 5". Each pair of supporting columns was linked by an 1" x 1½" semicircular shaped iron stay, bolted to the supports through forged eyes and, at the top centre, bolted to the timber cross member. This design of awning, first set out in 1903 on drawing No.6980, underwent various modifications as time went on, and became the basis of the awning used on the first 8 ton rollers derived from the *Reading* design. These were 8' 0½" long (or 8' 6½" for a roller with a Colonial firebox) by 4' 9" wide, but the materials and method of construction were the same save that cross braces were introduced between the stanchions at the tender end. A new design based upon the early 10 and 12 ton rollers, but with differences of detail, was made for No.25646, the first No.4 tractor which was ordered with an awning.

The roof on No.26063 for Jacob Studt (Fig.27) was what was subsequently described as an 'extended awning', reaching from the rear of the tender to just behind the chimney and carried on three pairs of columns, of which the front pair was carried on sockets and brackets rivetted to the boiler, the intermediates flattened at the foot and bolted to the hornplate, and the rear fixed to the back of the coal space. This last pair was cross braced and the centre pair had inclined struts to the rear end of the hornplates. The roof itself was very similar in construction to that fitted to No.25646 but with some obvious improvements. The three main roof members were of the same scantling (1½" x 12") though the drawing, No.8568, is silent as to the timber used. There were four intermediate ribs, which were described as 'ledges' on the drawing, 2" wide and ¾" thick following the curve of the roof, possibly steamed to shape out of ash, but again the drawing is silent. The width of the awning remained at 5' 0" but the length was 11' 8⅛" and, as before, the covering was ½" unmatched boarding and 24 gauge galvanized iron.

Subsequent tractors were roofed in a different style, the awning extending only to slightly forward of the crank box, on two pairs of stanchions, with the covering and boarding brought down in a continuous sweep, in Fowler fashion, to the bottom of the fascias. Originally 8' 5" long by 5' 0" wide, they were soon widened to 6' 0". The top was ½" deal boarding covered in canvas bedded in white lead paste and painted. In the days when paints were mixed on the premises the bedding paint contained the skins from the tops of paint tins and kettles ground up with residues from the bottoms of tins. The two cross members - there were no intermediate ribs - were 1½" x 12", initially of elm but later of pitch pine or, later still, poplar. When asked for an 'extended awning' the firm provided a roof in similar design but extended to the front of the cylinder, from which it was supported. Both of these types of roof were still being supplied to the end of the independent existence of Richard Garrett & Sons Ltd. One of the last batches to have extended awnings of this design was that for the Eddison Steam Rolling Co. in 1919, Nos.33670, 33738, 33739, 33784, and 33812. During its currency this type of awning had undergone a number of minor variations, one of which was to finish the ends of the roof so as to project in an elliptical form on plan. This was supplied in two variants. One extended forward to about the trunk guides, the first so fitted being No.28076 of 1910, the other to the crankshaft end of the cylinder block, not quite so long as the other form of extended awning and still carried on two pairs of stanchions rather than on three. An example of this latter was on No.32703 for Alban Thomas in 1915, which was afterwards owned by Robert Wynn & Sons of Newport. On tractors ordered with an 'extended awning' the roof was brought forward, on three pairs of stanchions, to just behind the chimney. Such a tractor was No.32358 (June, 1914) for W.J. Last, the haulage contractor of Stonham Aspal, Suffolk. At the end of the war the main ribs were made of poplar with 1" x 1" x 1/8" steel angles bolted to them against the underside of the roof boards. As can be seen from the illustrations the ends of the vertical sides were shaped to something like an ogee profile blending rather pleasingly into the elliptical shape of the awning ends. Initially 8' 5" long these awnings were changed to 9' 5" in August, 1910. At first the underside of the boarding was 6' 6" above the floor of the manstand, but this was dropped to 6' 3" in December, 1912. In the showman's tractor No.27342 supplied to Charles Hewson in 1909 the same basic pattern of awning was used, extended to 17' 0¼" overall so as to reach out over the dynamo bracket. Just above the rear of the crankbox an additional roof bearer 5" wide and 2½" deep was introduced to take the weight of the bracket carrying the extension chimney when lowered. At Hewson's request 9" deep nameboards were fitted outside the awning. In effect, these formed continuous rainwater gutters. The roofing canvas was dressed down into the trough and over the back and top of the upward projecting nameboards, to be secured a little way down the outer face. This saved water dripping down the sides of the engine but meant that it was discharged at each corner. The awning on Tommy Cottrell's showman's tractor No.31025 was essentially similar, though the fascias were a little narrower.

Similar designs of awnings were adopted for rollers but, because of the way in which matters of detail were left to the discretion of individual draughtsmen, the sizing of members was not identical. Thus the *Reading* and its lighter derivatives, when ordered to be so fitted, received awnings visually like that shown in Fig.137 and referred to in the preceding paragraphs. A very similar awning set out on drawing No.13309 (October, 1912) for the 3 ton tractor and the 4 ton roller derived from it showed, perversely and probably for no better reason than that another man drew it, differing ribs and eaves members. When contemplating these diversities of detail one is tempted to wonder how they were treated in the thresher department charged with the task of actually making the awnings. Could it be that 'shop practice' (i.e. common sense) ignored some of them? On the rare occasions when awnings were ordered on heavy rollers the original pattern was followed for several years, but latterly official practice was to use the design (drawing 10610) devised originally for the 6NHP road locomotives, again very much in Fowler style.

The tractors built for the Ministry of Munitions paved the way for the final design of awning used on the 4CDs after their revision and up-dating in 1919 from which they emerged rather stronger and heavier than before. In this pattern of awning the radius of curvature was decreased to 7' 2" which was constant across the roof. Earlier awnings had been curved on top to 9' 0" radius, and on the heavy rollers to as flat a curve as 12' 0 radius. The sharper curvature at the eaves was abolished, and the roof boards reduced to 3/8" thick deal match-boarding, lapped over 1" x 7" deal fascia boards. This was probably the neatest and most workmanlike of any of the awnings used. Notwithstanding the diversity of awnings that have been described there were variations of detail that I have not attempted to catalogue. These included such things as the shaping of the ends of the fascias and the like. Besides all this the use of various patterns of awnings overlapped one another, probably to meet customer preferences. It is difficult otherwise to think what might have prompted so many designs in a relatively short period. The impending relaxation of the unladen weight limit to 7¼ tons made the weight of components a less pressing matter in 1919, not that the earlier limit of 5 tons had been paid more than lip service. Only the painters' weights put onto the tractors reflected the legal requirement - a weigh-bridge would never have done so.

Other characteristics that altered over the period of building were the front tank and the rear draw-gear. The change from the deep tank, extending down to the bottom of the firebox and necessitating a side-opening ashpan damper, to the shallower pattern reaching nearer to the perch bracket has already been related and does not need to be restated here. The revision of the draw-gear, however, came somewhat later and derived from the requirements of the Ministry of Munitions. The flanged pin plate used up to that time, which had 3½" between the parallel top and bottom flanges, allowed sufficient latitude for coupling up to a thresher or a traction wagon where the design of front lock permitted upward or downward movement of the drawbar. Obviously someone at the Ministry had envisaged the problems in coupling up to a military load - perhaps an artillery weapon where the drawbar was heavier and rigid, as it would have been, indeed, with any two-wheeled trailer. In Fig.37 the old type plate can be seen clearly, whilst the revised type is visible in Fig.129 where it will be noticed that, though the flanges remain 3½" apart where the pin passes through, outside that limit they are flanged up and down at about 45° thereby allowing not only greater flexibility in the vertical plane but also making it much easier to couple up the drawbar of the towed vehicle.

The 4CD tractor was an inherently handsome design that suffered a good deal of indignity from the extras foisted upon it. For instance the use of superheating, whether as the earlier coil in the goitred chimney base or the later design in the flat-topped smokebox, did nothing for the appearance of the tractors. For that matter, nor did the addition of the optional feedwater heater either in its original form, as shown on drawing 13943 and fitted to Alfred Dawson's tractor No.28220, or in the revised form (drawing 15056), first used on superheated tractor No.32360 for William Aitken of Linlithgow, situated on the top of the boiler barrel just in front of the cylinder block. Somehow it never failed to give the impression of improvisation.

What practical effect the waterheater had upon the economy of a tractor to which it was fitted is, to say the least, problematical as seemingly no controlled trials were conducted. A measured test conducted upon both 3 ton and 5 ton wagons in March, 1919, concluded that the coal consumption of a loaded 5-tonner was decreased by 6.9% by the use of the feedwater heater. On the other hand the 3 ton wagon test showed no economy in the quantity of fuel used. The water heater added some three quarters of a hundredweight to a vehicle already seriously in excess of its legal weight. One may conclude that trials on steam tractors might have yielded similarly ambiguous results, but this remains no more than conjecture.

A mechanical lubricator for the cylinder was always an extra over the basic price of the roller or the tractor, and therefore, at least nominally, in the choice of the customer or agent. Although few of the 'heavy' rollers had more than a displacement lubricator, a 'bought-out' component in any event, a number were provided with Roscoe lubricators, an improved type. On the rare occasions when a mechanical lubricator was asked for the Mollerup ¼-pint size was often fitted, as in the cases, picking at random, of Nos.26567, 27323, and 27347, respectively for Reifenrath in Germany, one of Roll's customers in Holland, and another in Dutch East Indies. The single cylindered No.3 and 4 tractors also were fitted merely with a displacement lubricator probably of no more than 3 fluid ounce capacity. With the 4CD tractor, however, the pattern was rather different. The very first to be delivered, No.26063, had a 1-pint Manzel mechanical lubricator. Though I cannot claim to have checked percentagewise on the cylinder lubricators fitted to 4CDs the Manzel had a very substantial share, as did the Vacuum Oil Co's. 'Pony' size lubricator. From time to time others cropped up such as the Empire 'Trusty'. By 1907, when production of the 4CDs began, the Mollerup lubricator was, in any case, fast approaching its sell-by date, and its use tapered off to extinction by about 1911. The last rollers which I have noted as being fitted with Mollerup lubricators were Nos.29021 and 29022 delivered to Strassenwalzen Betrieb (successors to Reifenrath) at Niederlahnstein, Germany, in May and June, 1911.

Perhaps the most striking change in the external appearance of the 4CDs, to the casual observer at least, was the dropping, c.1919, of the circular brass plate, bearing the legend 'The Garrett Tractor', from the smokebox door and the substitution of the single large letter '**G**' cast in brass. This led some to refer to the last series of tractors as the 'Big **G**s'.

These final tractors and their predecessors earned themselves a high regard with their owners and users both for their dependability and their capacity for work. They were, besides, well liked for their quietness in running and their consistent performance on belt work because they governed well. Whilst no machine can be expected to be a paragon of all the mechanical virtues the 4CD tractors came out at or near the top of their league. Indeed the only real criticism of them that I have heard was that they could not quite keep up with their Aveling & Porter counterparts in road speed. In the period of the 5 ton weight limit, of course, the 4CD exceeded it unblushingly but then so did the products of most other makers. A Garrett 4CD equipped to a showman's specification with a dynamo and bracket, full length awning, extension chimney, and decorative brasswork probably topped 9½ tons without coal and water but the firm seems to have caused such an engine to keep carefully away from the weighbridge. To have made this, the concluding volume of my account of Garretts' self-moving engines, mainly a record of their tough and handy tractors seems an appropriate way in which to bring the series to a conclusion.

Fig.133. *September, 1949, at Red House Farm, Ufford, and 4CD No.32860 (new in June, 1916), owned by H.J.Holland & Sons, is tackling, to judge from the rig over the feeding hole, a stack of beans.* [R.G.Pratt]

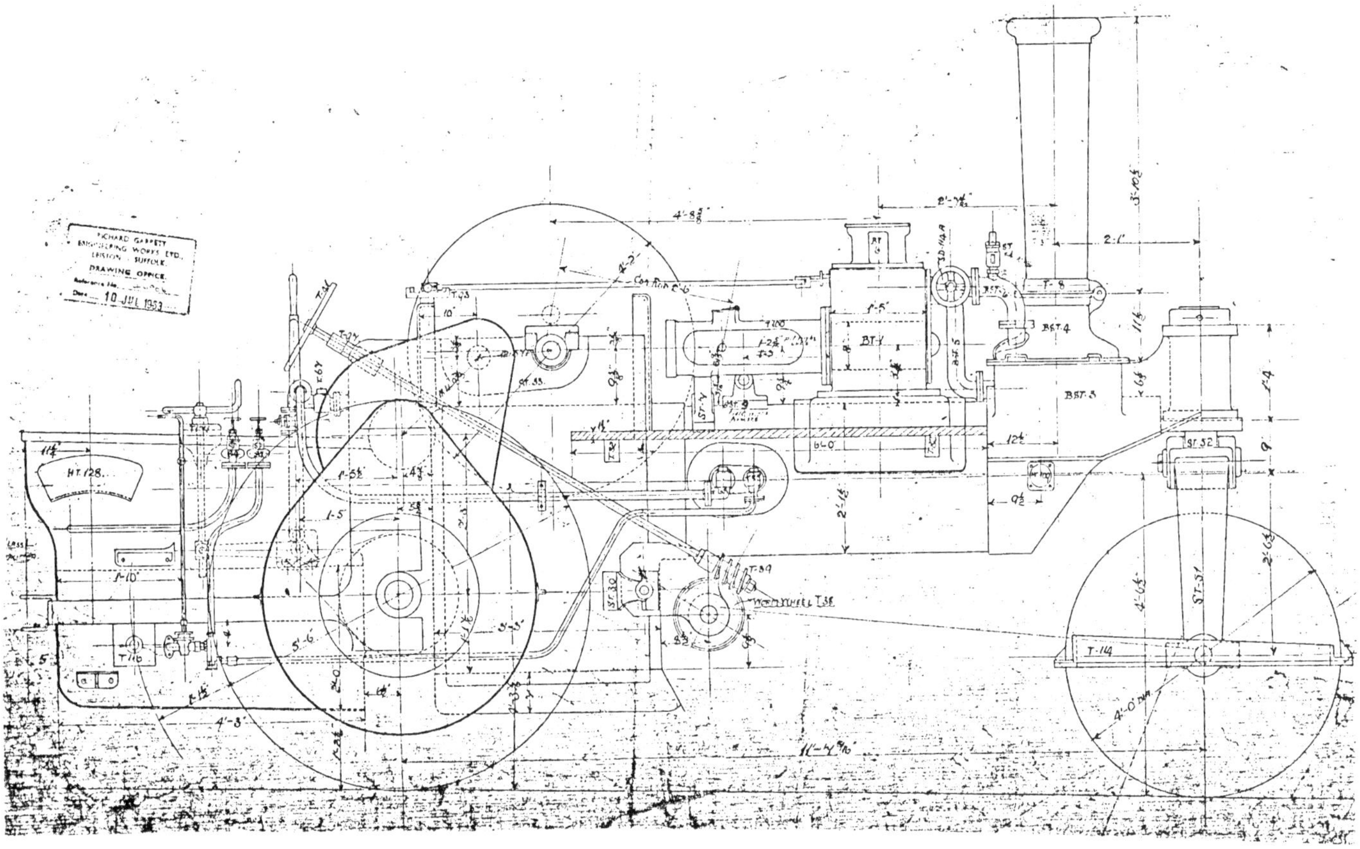

***Fig.134.** Gear-side elevation of the10 and 12 ton superheated single cylinder four-shaft roller.*

***Fig.135.** The 5 ton single cylinder three-shaft roller, code-named* Earl*, which used the 10" stroke* Reading *cylinder block bored to 6½".*

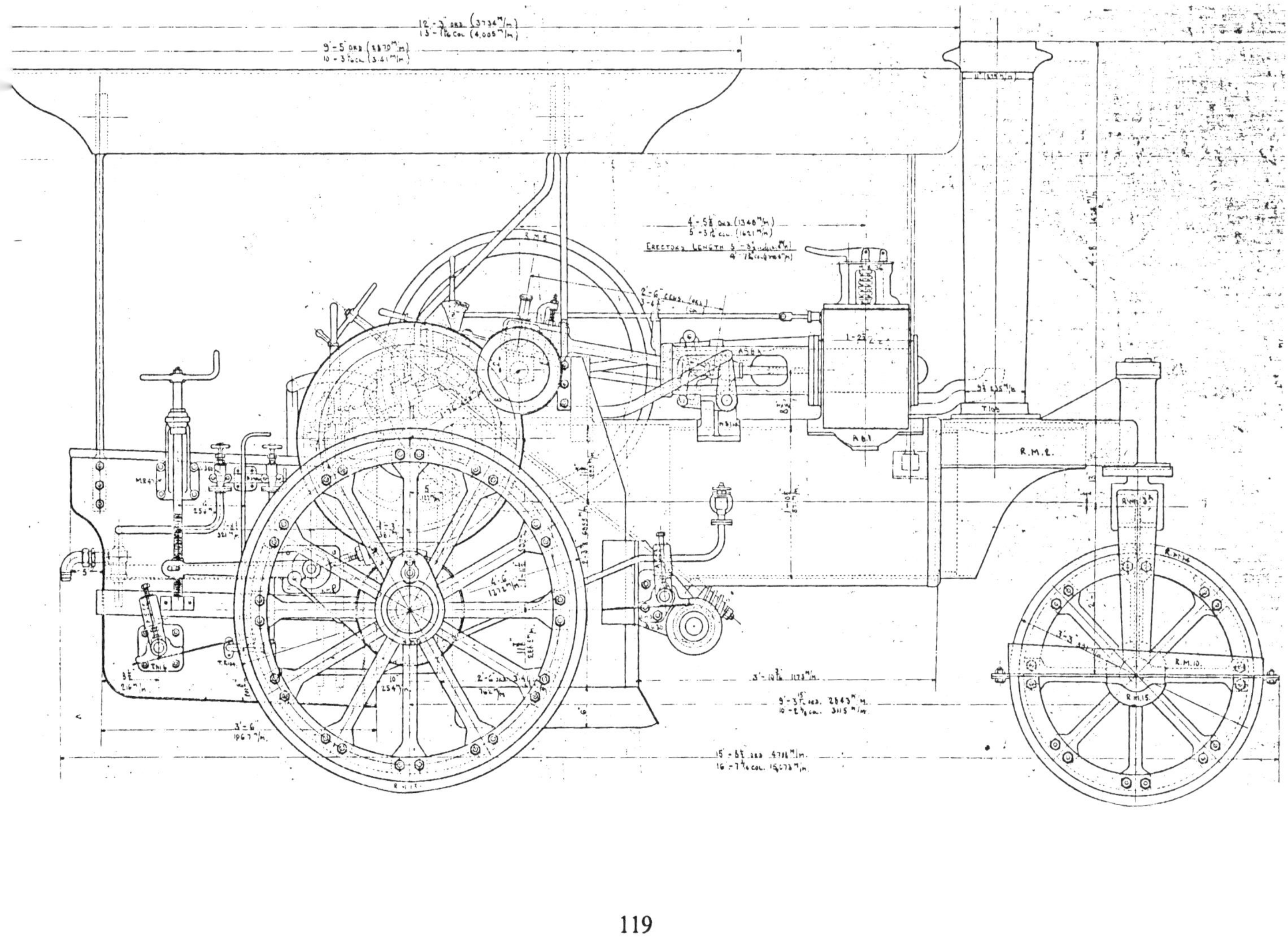

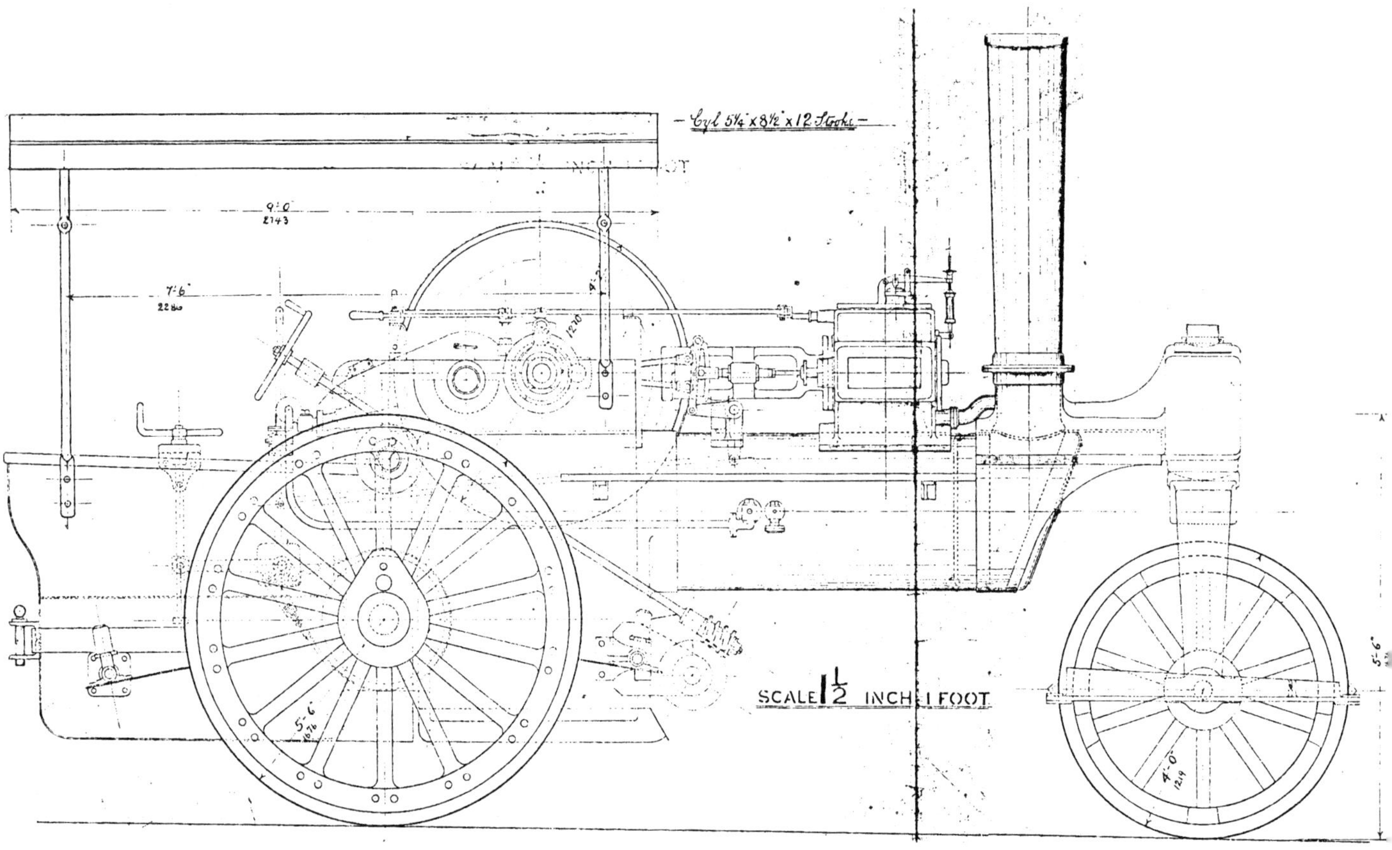

Fig.136. *The 10 ton compound roller as originally drawn in 1903.*

Fig.137. *The 4 ton single cylinder roller No.30268 (June, 1912), based upon the No.3 tractor, supplied to the Ministerio Obras Publicas in Buenos Aires, via Mattos. It had a Colonial firebox and grooved rolls sealed for water ballasting. The grooves were formed by wrought iron rings shrunk on and secured by studs drilled and tapped into the rims. The Pickering governor suggests it was meant also to work on the belt.*

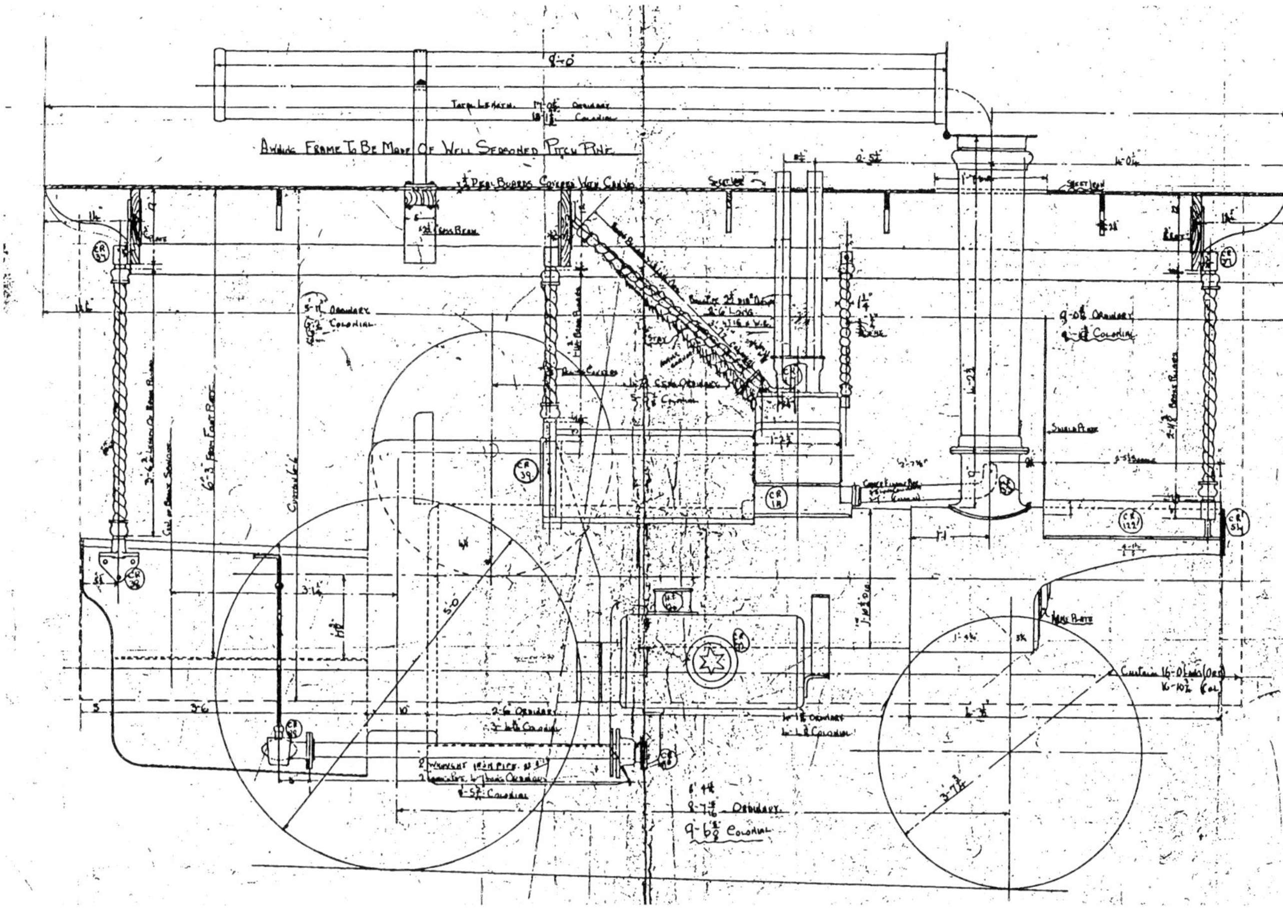

Fig.138. The arrangement of the showman's tractor, 1909, showing the deviations from standard plus details of the roof as used on Chas. Hewson's No.27342.

Fig.139. The gearside elevation of the 15 ton compound roller as re-drawn 1923/24. None were built to this design at Leiston, but it was used in the Ansaldo-Garrett rollers.

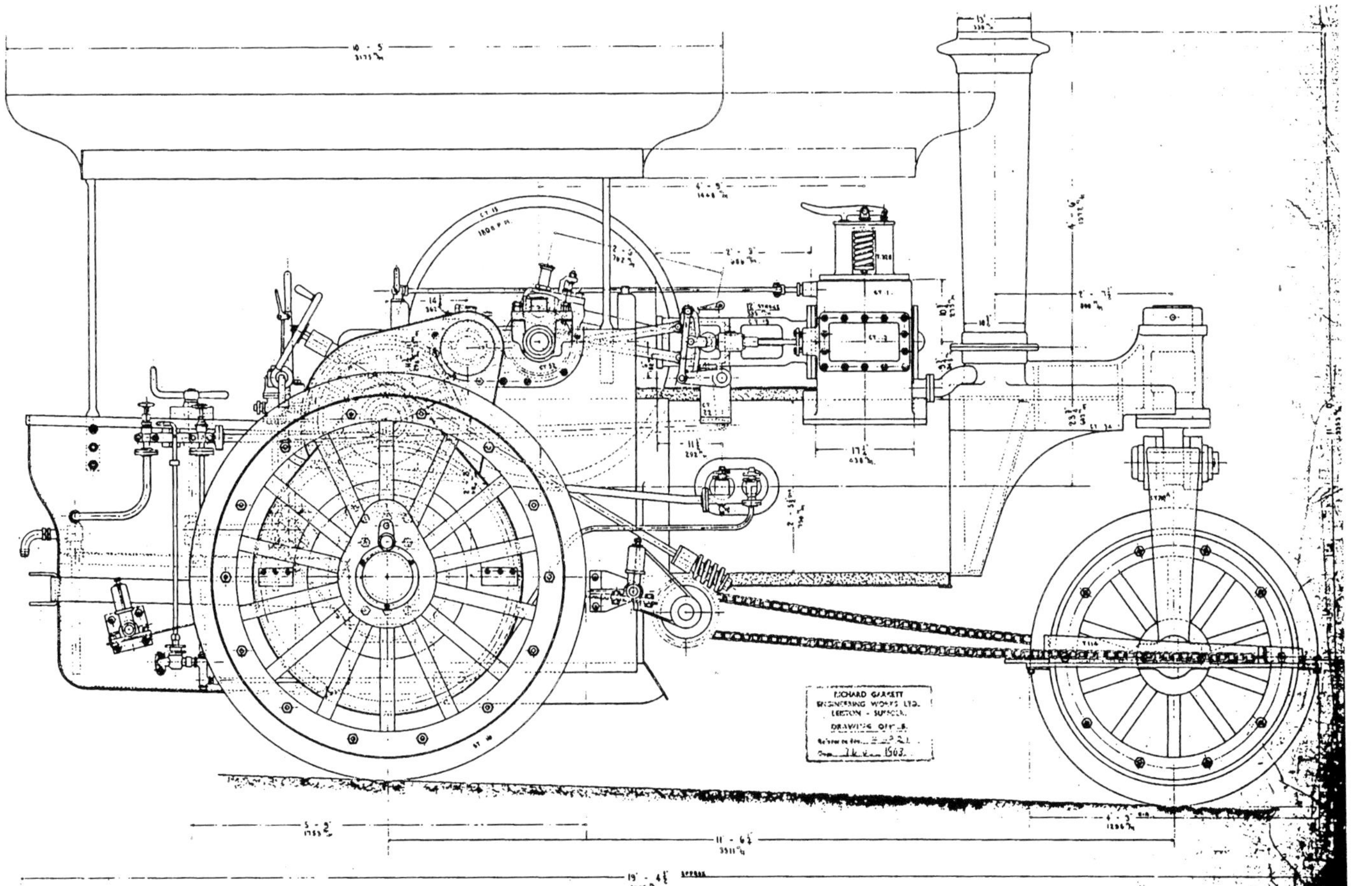

Fig.140. *Even good tractors are sometimes faced with circumstances they cannot control. In this picture No.31751, of the Norfolk Motor Transport Co. Ltd, is shown on its side at Earlham Bridge, Norwich, in 1915, after having been pushed off the road and down a bank by a load of beech logs. Although the tractor was fitted with the gear to enable the driver to apply the trailer brakes from the manstand this did not prevent the load taking control in this instance, possibly because it was out of order or, dare one say it, because it was too much bother to connect it up at all.*

In ***Fig.141.*** *it is back on its rear wheels, having had its front axle and wheels detached for ease of recovery, after having been lifted by the Aveling & Porter crane engine of J. & H. Bowles, also of Norwich. The tractor was not seriously damaged and was soon back at work.*

Fig.142. No.33981 first went to W.H.Gaze & Son of Kingston, Surrey, in 1920 but was returned and re-sold as second-hand in October, 1921, to R.N.Munn of Hunton, Kent. Chris Lambert bought it in 1925 and it was photographed in his yard at Horsmonden in May, 1946 by Derek Stoyel. It is preserved.

Fig.143. The previous month Derek had recorded No.33705 in the same yard. New as a tractor to Stone Court Brick & Tile Works at nearby Pembury it was bought third-hand by Chris in 1931, going on to Syd Sutton at Beckermet in 1949. It was converted to a showman's tractor in preservation, and is now in Japan. The roller in the background is No.34461, also a converted tractor.

Fig.144. An unidentified but later Garrett single cylindered 10 ton roller owned by R. Dingle & Sons at Stoke Climsland, Devon.
[R.Sandercock collection]

Fig.145. *Twelve ton single No.31295* One and All, *(Dec.1912), owned by R.Dingle, equipped with the earlier pattern of forecarriage head and photographed in the 1920s.* [R.Sandercock collection]

Fig.146. *A youthful Billy Mayne posed by the ex-Ministry of Munitions tractor* Nippy, *No.33325, owned by his father J.Mayne. After the Maynes went over to motors the tractor continued in fairground use with John Griggs of Petersfield, and ended its days at Mitcham with Tom Reynolds.* [J.B.A.Irvin collection]

Fig.147. *John Cole's* Queen of Great Britain *(No.33486 of April, 1919) set up for work at The Quay at Bath, Somerset. New to Browning Bros. of Stonehouse, Glos. it went to John Cole in 1923. It is now in preservation.* [A.J.Martin collection]

Specification of the "GARRETT" Standard Steam Road Rollers.

General Description. These Engines are constructed on the four-shaft principle.

Steel has been substituted for iron throughout wherever it has been found practicable to do so.

Every practical device for ensuring economy of fuel and water, without increasing the complication of wearing parts and thus adding to the cost of upkeep, has been adopted.

Special care has been given to the question of the proportion of Engine to boiler, and the Garrett Steam Road Rollers are mounted on boilers (with the **Garrett Corrugated Firebox**) of such sizes as to be capable of supplying steam in the driest possible condition well in excess of the maximum requirements of the Engine mounted upon it without being unduly forced; to this must be attributed in some measure the unrivalled reputation which the Garrett Steam Road Rollers enjoy for **economy of fuel, easy steaming, and low cost of upkeep.**

Cyiinder. Of best cold blast, close-grained cast-iron, steam jacketted and surmounted by steam dome wherein are placed the safety valves and starting valve, ensuring the use of dry steam. Lagged with approved non-conducting material, covered with sheet steel casing, and provided with all necessary fittings and lubricators of best quality.

Note.—The Cylinders of the Garrett Compound Steam Road Rollers are constructed with **outside value chests, making the slide valves easily accessible,** and are provided with an auxiliary valve to admit high pressure steam into low pressure cylinder when starting engine; this auxiliary valve is operated from footplate, and arranged to close automatically.

Piston. Accurately fitted with rings of most approved pattern. Piston rods of best quality mild steel.

Crosshead Guide. This is of the trunk type cast in one piece with the front cylinder cover. It is bored and faced at one setting so that perfect alignment is obtained.

Crosshead. This is of cast steel fitted with special mixture cast-iron slippers of wide bearing surface accurately machined to fit inside the trunk guide and adjustable so that wear can be taken up.

The gudgeon, through which the small end of conecting rod is secured, is of large diameter, deeply case-hardened and ground to an exact fit and has tapered seatings.

Connecting Rod. Forged out of a hammered ingot of finest quality steel, machined to limit gauge throughout and polished; the large end of connecting rod is of exceptionally substantial design, the strap being bolted to the rod and the bearings adjusted by steel cotters.

Crankshaft. The Garrett Standard Single Cylinder Steam Road Rollers have crankshafts forged and bent out of solid hammered forgings of finest steel.

The Garrett Standard Compound Steam Road Rollers have double throw, right angle type crankshaft made from solid forgings of finest steel, the webs being machined out of the solid according to the best locomotive practice and accurately balanced with heavy balance weights securely attached to same on the most approved method.

Countershafts and Main Axle. The two countershafts and main axle are forged from hammered ingots of the finest steel and machined to limit gauges. The first countershaft is machined "square" where the "change-speed" double pinion slides, so that all loose keys are dispensed with.

Specification of the Garrett Standard Steam Road Rollers—continued.

Gearing. All the gearing is of the best crucible cast steel to a special analysis which long experience has proved to be the most durable.

All the fast running gear wheels (*i.e.*. for the first and second transmissions) are "**machine-cut**" by our own special machinery with the most **mathematical accuracy,** with the result that **all vibration,** usually set up by fast-running gear wheels with the old-fashioned cast teeth or cogs, is **entirely avoided,** thus **reducing the wear and tear throughout the whole Engine and boiler to the lowest possible point** as well as **rendering the road roller noiseless in working.**

The gearing is arranged for **two speeds** and in such a way that only one speed can be put into gear at one time; it can moreover be easily disconnected to enable a stonebreaker or other fixed machinery to be driven by a belt off the flywheel of the roller.

The whole of the outer gearing is enclosed in stout flanged steel casing.

Reversing Gear. Link motion reversing gear of locomotive type with double powerful lifting levers and made of best mild steel throughout. The joints and wearing surfaces are extra large, deeply case-hardened and ground to an exact fit.

Flywheel. Heavy cast-iron flywheel of disc type ensuring steady running of the Engine, turned and polished on the face. All sizes of the Standard Garrett Steam Road Rollers from and including the 10-ton size and upwards are fitted with flywheels having faces of suitable width and correctly cambered to take belt. The flywheels of the 6 and 8 ton sizes of the Standard Garrett Steam Road Rollers are not wide enough on the face to take belt, and if required also for driving fixed machinery at any time, a small pulley (18in. diameter by 6in. wide being maximum size) may be ordered specially at extra charge to be fitted on crankshaft on opposite end to flywheel.

Bearings. All bearings and bushes are of best quality gun-metal and of exceptional width to ensure **cool running;** they are fitted into recesses bored in the hornplates so that all strains are taken by the plates, which are specially stayed up to receive these strains.

Steerage. Of the worm and worm-wheel type.

Special Conveniences for Driving. The starting and reversing levers, "change-speed" gear lever, feed-pump, brake handle and steerage wheels are all placed in convenient positions round the driver's stand so as all to be within easy reach of the driver, thus enabling one man to both drive and steer.

Lubrication. The greatest possible care has been given to the question of providing efficient lubrication for all the moving parts of the Road Roller and especially for those bearings and moving parts which cannot be attended to when the Road Roller is in motion.

Brake. A powerful friction band brake, acting on drum keyed to main axle, is provided.

Hind and Front Rollers. The rims or tyres are of a special mixture of hard cast iron, both the hind and front rims being renewable, and the front rollers are fitted with renewable bushes.

Scrapers. The hind rollers are fitted with our latest improved type adjustable spring scraper, both in front and behind: the front rollers are provided with adjustable scrapers consisting of steel angle bars.

Feedpump. Feedpump of ample capacity fitted inside hornplates and driven by separate eccentric off crankshaft, provided with all necessary pipe connections, clack box, &c. A special cock is also provided between clack box and boiler by means of which the pump can be entirely disconnected from the boiler. This is most convenient should dirt or grit get under the pump valves, necessitating their being taken out and cleaned whilst the Engine is under steam.

Injector of automatic re-starting type (Penberthy or other improved make) with all necessary pipe connections, cocks, check valves, &c. A special cock is provided between the valve box and boiler by means of which the injector can be entirely disconnected from the boiler at will. This is most convenient should dirt or grit get under valve, necessitating its being taken out and cleaned whilst the Engine is under steam.

Water Lifter. A powerful steam water lifter, with all necessary cocks, piping, and 26ft. of special wire bound suction hose, is provided.

Boiler. The boiler is designed and constructed throughout according to the requirements of the British Board of Trade, the Manchester Steam Users' Association, and all first-class Boiler Insurance Companies, and will pass their respective tests for the working steam pressure for which they are sold, being of the same strength in proportion and workmanship as adopted in the best locomotive practice.

Specification of the Garrett Standard Steam Road Rollers—continued.

Boiler—*continued.*

Siemens-Martin mild steel plates, properly annealed, are used throughout—the shell plates, internal firebox plates and round tube plates being of the respective qualities specified by the British Board of Trade.

All plates are specially tested and certified by Lloyd's Surveyor for Richard Garrett & Sons, Ltd., before leaving the makers' works.

All edges are planed or milled and properly fullered with a broad tool after rivetting.

All holes for rivets and stays, where possible, are drilled in position after the boiler has been plated together.

Rivetting is done by the most modern hydraulic machinery available.

Flanging is done by powerful hydraulic flanging presses **at one heat.**

Outer faces of flanges to round tube-plates are machined to ensure perfect fullering.

The hornplates are of extra strong steel plates and **are securely stayed together by two specially designed, stout, transverse stay-plates fitted in between them and securely bolted to them; these transverse stay-plates are flanged over at the top as well as round at the sides,** renders it practically impossible for them to bend, and thus preventing vibration of the hornplates.

Longitudinal seams are double rivetted, also the circumferential seam joining firebox shell to barrel.

Solid foundation rings are fitted to all sizes of the Garrett Steam Road Rollers and angle iron is entirely dispensed with.

A number of specially stout **stay-tubes** threaded at both ends are securely screwed into both tube-plates; these are in accordance with the British Board of Trade's highest class regulations and greatly increase the strength of the boiler.

The **other tubes,** which are expanded into both tube-plates according to the best practice, are of **finest seamless steel, solid drawn.**

Boiler barrel of **extra large diameter,** giving an **unusually large steam space,** and consequently ensuring the use of **dry steam.**

Boiler Cleaning Facilities. A large manhole is placed in the side of the boiler barrel within easy reach of the firebox, giving every facility for internal examination and cleaning.

Mudholes are provided at all four corners of the firebox just above the foundation ring making the water spaces on all four sides easy to clean out; these mudholes are so placed that the bridges and covers can be easily removed for cleaning and replaced after cleaning.

Further mudholes are provided at the top and bottom of smokebox tube-plate and in the front plate level with the firebox crown.

Boiler Fittings. A complete set of steam fittings of the most substantial and approved pattern is supplied as follows:—Two water gauges with stout brass guards; pressure gauge; whistle; blow-off cock; double safety valves; blast pipe and cock; special gland packed steam cocks for injector and water-lifter; check valves or clack boxes for injector and pump provided with special cocks between the clack boxes and the boiler, whereby the boiler may be completely disconnected when under steam; fusible plug, &c., &c.

Firebox. The **GARRETT CORRUGATED FIREBOX**

of

UNEQUALLED STEAMING CAPACITY, ECONOMY, and DURABILITY.

The **special mudhole** fitted in the external firebox front-plate, over the firehole, between the longitudinal stays, gives **easy access** to the **Garrett Corrugated Crownplate;** and there are **no rows of roof stays,** as with a flat topped firebox, to make difficult the task of cleaning the crownplate.

Tests (Boiler and Engine). The Boilers of the Garrett Standard Single Cylinder Steam Road Rollers are tested by **hydraulic pressure** (cold water) to **280 lbs. per square inch** for a regular **working steam pressure of 140 lbs. per square inch.**

The Boilers of the Garrett Standard Compound Steam Road Rollers are tested by **hydraulic pressure** (cold water) to **320 lbs. per square inch** for a regular **working steam pressure of 160 lbs. per square inch.**

Specification of the Garrett Standard Steam Road Rollers—continued.

Tests (Boiler and Engine)—*continued.*

Every Garrett Steam Road Roller has to undergo a prolonged brake trial at its regular working steam pressure, carrying full load and lasting several hours, when indicator diagrams are taken; no Engine is allowed to leave Leiston Works without having passed through this trial satisfactorily.

In addition to the above tests every Garrett Steam Road Roller is also tested over a variety of roads.

Tender. The Tender is constructed of extra stout mild steel plates with capacious water tank underneath footplate and coal bunker on the top; all corners are flanged by special hydraulic machinery with large "rounding" curves instead of sharp corners so as to assist cleaning and prevent accumulation of mud and dirt, &c.

Ashpan and Gear Covers are of stout steel plates flanged by special hydraulic machinery with large "rounding" curves instead of sharp corners to prevent the accumulation of dirt, &c., and to assist cleaning.

Special Drawbar Attachment so arranged that the draught is taken direct by the main axle without passing through the tender.

Guard Plates to Engine Motion Work. Engine motion work neatly enclosed by strong steel casing, brass beaded at the edges.

Lagging. Boiler lagged with wood, neatly covered with sheet steel casing, and brass bands with straining screws are fitted round barrel.

Painting. The Engine is painted, lined out, and varnished in best locomotive style.

Outfit. With every Garrett Steam Road Roller the following outfit is provided:—Waterproof cover of best Willesden or equal cloth; one pair large head lamps; one tail lamp with sliding shutters; one water gauge lamp for night work; toolbox with all necessary spanners, hammer and caulking tool; oil can; fire irons; tube brush; one set of frost pegs for hind rollers.

Extras. The following (when ordered specially) can be fitted at extra charge (prices on application):—

High Speed Pickering Type Governor of latest improved design.
Awning, either (1) over Tender, or (2) over full length of Engine, as preferred.
Differential or compensating gear.
Special large Firebox for wood, &c. (of the Garrett Corrugated Type).
Oil burning apparatus.
Cast steel rims instead of standard cast-iron for front or hind rollers.
Winding drum and 50 yards of steel wire rope.
Scarifier.
Close packing for shipment.

Specification of the Garrett Standard Steam Motor Tractor.

General Description. These Engines are constructed on the three-shaft principle. Steel has been substituted for iron throughout wherever it has been found practicable to do so.

Every practical device for ensuring economy of fuel and water, without increasing the complication of wearing parts and thus adding to the cost of upkeep, has been adopted.

Special care has been given to the question of the proportion of Engine to boiler, and the Garrett Steam Motor Tractors are mounted on boilers (fitted with the **Garrett Corrugated Firebox**—see pages 4 and 5) of such sizes as to be capable of supplying steam in the driest possible condition well in excess of the maximum requirements of the Engines mounted upon them without being unduly forced: to this must be attributed in some measure the unrivalled reputation which the Garrett Steam Motor Tractors enjoy for **economy of fuel, unequalled steaming and hauling qualities and low cost of upkeep.**

Cylinders. Of a special mixture of best cold-blast, close-grained cast-iron, steam jacketted and surmounted by steam dome, wherein are placed the double safety valves and starting valve, ensuring the use of dry steam.

Lagged with approved non-conducting material and covered with sheet steel casing.

Provided with all necessary fittings and lubricators of the most substantial and approved design, and of first-class quality throughout.

Special Note.—The cylinders of the Garrett Compound Steam Motor Tractors are constructed with **outside valve chests making the slides easily accessible,** and are provided with an auxiliary valve to admit high pressure steam into low pressure cylinder when starting Engine: this auxiliary valve is operated from the footplate, and arranged to close automatically.

Pistons. Accurately fitted with rings and springs of most approved pattern. The piston rods are of best quality mild steel.

Crosshead Guides. These are of the trunk type, cast in one piece with the front cylinder covers. They are bored and faced at one setting so that perfect alignment is obtained.

Crossheads. These are of finest cast steel fitted with slippers of special mixture cast iron and of specially large bearing surface, being accurately machined to fit inside the trunk guide and adjustable so that wear can be taken up.

The gudgeons, through which the small ends of connecting rods are secured, are of specially large diameter, deeply case-hardened and ground to an exact fit and have tapered seatings.

Connecting Rods. Forged out of hammered ingots of finest quality steel machined to limit gauge and polished: the large ends of connecting rods are of exceptionally substantial design, the straps being bolted to the rods and the bearings being adjusted by steel cotters.

Crankshaft. The single-throw crankshafts of the Garrett Single Cylinder Steam Motor Tractors and the double-throw, right-angle type crankshafts of the Garrett Compound Steam Motor Tractors are both made from solid forgings of finest steel, the webs being machined out of the solid, according to the best locomotive practice and accurately balanced with heavy balance weights securely attached to same on the most approved method.

The crankshaft is machined "square," where the "change-speed" pinions slide, so that all loose keys are dispensed with.

Countershaft and Main Axle. The countershaft and main axle are of the largest practical diameter, and are forged from hammered ingots of finest quality steel, and machined to limit gauge.

Gearing. All the gearing is of special width and of the best crucible cast steel to a special analysis, which long experience has proved to be most durable.

All the fast running gear wheels (*i.e.* for the first and second transmissions) **are machine-cut** by our own special machinery with the most **mathematical accuracy,** with the result that **all the vibration,** usually set up by fast running gear wheels with the old-fashioned "cast" teeth or cogs, **is entirely avoided,** thus **reducing the wear and tear throughout the whole Engine and boiler to the lowest possible point as well as rendering the Engine noiseless when travelling on the road.**

The gearing is arranged for two speeds, and in such a way that only one speed can be put into gear or partly into gear at one time: the gearing can be easily entirely disconnected to enable suitable fixed machinery to be driven by a belt from the flywheel of the Engine.

The whole of the outer gearing is enclosed in extra stout sheet steel casing.

Differential or Compensating Gear, consisting of two bevel pinions of best crucible cast steel having shrouded teeth or cogs (the shrouds being accurately faced by special machinery) and so arranged that they cannot gear too deeply.

The action of the differential gear is automatic and allows the hind driving wheels to revolve at different speeds or even in opposite directions, whereby all grinding action on the road surface with the consequent undue strain on the driving wheels, main axle, bearings and gearing, &c., is avoided when turning even the sharpest corners at a high rate of speed.

Reversing Gear. Link motion reversing gear of the most approved and up-to-date locomotive type with double powerful lifting levers and made of best Yorkshire or equal quality iron throughout.

The joints and wearing surfaces are of specially increased width, deeply case-hardened and ground to an exact fit.

Flywheel. Heavy cast-iron flywheel of the disc type having rim of special thickness and weight to ensure steady running.

The face of the rim is turned to correct radius to take belt for driving suitable fixed machinery.

Steel Castings. All brackets and other castings subjected to heavy strains, such as the main axle brackets, intermediate or countershaft bracket, turnplate casting, winding drum, &c., are of best crucible cast steel whereby increased strength with reduced weight is obtained.

Bearings. All bearings and bushes are of best quality gun-metal and specially wide, ensuring cool running and longer life: they are fitted into recesses bored in the hornplates, so that all strains are taken by the plates which are specially stayed to receive these strains.

Steerage. Of the worm and worm-wheel type.

Special Conveniences for Driving. The starting and reversing levers, change-speed levers, feedpump, injector and water lifter gland-packed steam cocks, brake handle, steerage wheel, are all placed in convenient positions round the driver's stand so as all to be within easy reach of the driver, thus enabling one man to both drive and steer.

Lubrication. The greatest possible care has been given to the question of providing efficient lubrication for all moving parts of the Engine and especially for those bearings and moving parts which cannot be attended to when the motor tractor is travelling on the road.

The lubrication of the cylinders is effected by means of an extremely reliable and durable type of positive feed oil-pump (patent).

Specification of the Garrett Standard Steam Motor Tractor *continued.*

Travelling Wheels. Both the front and hind driving wheels are of the largest practical diameter and are of specially strong construction throughout.

The tee rings of the hind driving wheels are rolled with specially strong section having wide cross-strips or strakes rivetted on the faces in accordance with the requirements of the "Heavy Motor Act": these cross-strips are supplied of steel or iron as preferred and have hard wood keys driven in (dovetail fashion) between each pair of cross-strips, which gives the very best possible grip on the road and is also unequalled for wear.

The hind driving wheels can be supplied to special order with tyres consisting entirely of wood blocks on a special system.

The spokes are of special design giving greatly increased strength, and the left-hand driving wheel is provided with two driving pins.

Brake ("Rim" Type). A powerful screw friction brake is fitted, operating on angle rings of strong section bolted to the insides of both hind driving wheels, and ensuring complete control of the Engine on the steepest hills: the brake handle is placed within convenient reach of the driver's hand on footplate.

Spring Gear. The Garrett Standard Steam Motor Tractors are mounted upon springs to both axles.

The front axle is mounted upon a single locomotive type laminated spring of special quality steel, and of the greatest possible strength consistent with efficient "springing action."

The hind axle is mounted upon two locomotive type laminated springs of special quality steel and of the greatest possible strength consistent with efficient "springing action": these twin springs being placed parallel to each other across the Engine.

The cast steel hind axle guide brackets (in which the cast steel axle boxes slide up and down with the action of the springs) have strong flanges underneath which rest upon the ends of the two springs.

The springs are held across their centres in a massive cast steel bridle through the centre of which the adjustment rod passes: this adjustment rod is of great strength and has a jaw at the upper end through which the pin passes which secures the connecting lever and round which the connecting lever can swivel vertically: the two ends of the connecting lever are secured to the two sliding axle boxes by stout steel pins.

The lower end of the adjustment rod which passes through the bridle is threaded and fitted with lock nuts by means of which the tension of the springs can be regulated to a nicety with the assistance of the spring indicator which is fitted to one of the sliding axle boxes inside the hornplate and shows the tension of the springs at any given time.

Three-eighths of an inch "springing movement" is allowed before the sliding axle boxes come to rest in the guide brackets, and this amount of "springing movement" has been proved by practical experience to be the utmost possible consistent with freedom from excessive cost of upkeep for the spring gear.

The teeth of the main driving spur wheel and pinion are specially deep to allow of a sufficient vertical sliding in and out motion corresponding with the action of the springs.

This system, which is recognised by users as giving all the "springing action" required, enables all the shafts to be fixtures except the main axle and consequently entirely does away with the complicated mechanism for keeping the counter-shaft and main axle the same distance apart, a mechanism which has been conclusively proved by users to be peculiarly liable to get out of order and thus to be very expensive in the matter of upkeep.

Winding Drum. A fast and loose type winding drum is fitted to the main axle, which can be easily thrown in and out of gear, and is provided with 50 yards of steel wire rope of special quality.

Feedpump. A special feedpump of ample capacity is fitted, suitably **geared down so as to prevent the pump running at an excessive speed** at any time.

The pump is fitted inside the hornplates and driven by separate eccentric off its own countershaft and provided with all necessary pipe connections, clack box, &c.

A special cock is also provided between the clack box and the boiler by means of which the pump can be entirely disconnected from the boiler: this is most convenient for examination or cleaning of the pump valves whilst the Engine is under steam.

Injector. An automatic, re-starting, high duty injector of most approved design is fitted, having special gland-packed steam cock as well as the handle of the water valve both placed in convenient positions to the driver's hand.

The injector is provided with all necessary pipe connections, also clack box, &c., whilst between the clack box and the boiler a special cock is fitted by means of which the injector can be entirely disconnected from the boiler, which is most convenient for examination or cleaning of the clack box valve whilst the Engine is under steam.

Water Lifter. Steam water lifter of extra large capacity with all necessary cocks (including special gland-packed steam cock) and pipe connections together with 26 feet of special wire-bound suction hose.

Boiler. In the design of the boilers of the Garrett Steam Motor Tractors, the greatest possible care has been taken to ensure an ample margin of strength in every smallest detail, so that the various severe strains, to which the boilers of this class of Engine are subjected, are amply provided for.

The boiler is constructed throughout according to the highest requirements of the British Board of Trade, the Manchester Steam Users' Association, and all first-class Boiler Insurance Companies, and will pass with a large margin to spare their respective tests for a regular working steam pressure of 180 lbs. per square inch, being of the same strength and workmanship as adopted in the best locomotive practice.

Siemens-Martin mild steel plates, annealed with the utmost care, are used throughout, the shell plates and tube-plate, and the internal firebox plates being of the respective qualities specified by the British Board of Trade.

All plates are specially tested and certified by Lloyd's Surveyor for Richard Garrett & Sons, Ltd., before leaving the makers' works.

The internal firebox and all flanged plates have a tensile strength of from 26 to 30 tons per square inch, with an elongation of not less than 25% in a length of 8 inches: the shell plates have a tensile strength of from 28 to 32 tons per square inch with an elongation of not less than 20% in a length of 8 inches.

All edges are planed or milled and properly fullered with a broad tool after rivetting.

Special attention is paid to the drilling of the plates—all holes for rivets and stays where possible being drilled in position after the boiler has been plated together—and to the rivetting, which is done by the most modern hydraulic machinery available, the rivets being of a special quality mild steel. The holes for the firebox screwed stays are drilled and tapped by the most modern machinery available to special gauges, and the screwed stays are of specially selected mild steel as specified by the Board of Trade.

Flanging is done by powerful hydraulic flanging presses **at one heat.**

Outer faces of flanges to round tube-plates are machined to ensure perfect fullering after being rivetted in.

The hornplates are of specially strong steel plates and **are rigidly stayed together by two specially designed transverse stay-plates,** specially strong, being **fitted in between the hornplates and securely bolted to them: one of these transverse stay-plates is flanged over at the top as well as round at the sides,** rendering it practically impossible for it to bend, and **thus preventing vibration of the hornplates.**

The cannon bracket, carrying the bearings for the intermediate shaft, is of cast steel and in one piece, and reaches from hornplate to hornplate, and is accurately fitted and securely bolted to them—thus providing a most valuable cross-stay to the hornplates in a place where one is most required.

The longitudinal seams are double rivetted, also the circumferential seam joining firebox shell to barrel.

Solid foundation rings are fitted and angle iron is entirely dispensed with.

Specification of the Garrett Standard Steam Motor Tractor—*continued.*

Boiler—*continued*

A number of specially stout **stay-tubes,** threaded at both ends by special machinery to special gauges, are securely screwed into both tube-plates, these being in accordance with the British Board of Trade's highest class regulations and greatly increasing the strength of the boiler: the tube-holes for these stay-tubes are threaded by special machinery with a continuous tap to special gauges.

The other tubes which are expanded into both tube-plates according to the very best practice are of finest quality seamless steel, solid drawn.

The boiler barrel is of specially large diameter, giving an unusually large steam space and consequently preventing priming by ensuring the use of dry steam.

Boiler Cleaning Facilities. A large manhole is placed in the side of the boiler barrel within easy reach of the firebox, giving every facility for internal examination of the boiler and cleaning.

Mudholes are provided at all four corners of the firebox just above the foundation ring, making the water spaces on all four sides easy to clean out; these mudholes are so placed that the bridges and covers can be easily removed for cleaning and replaced after cleaning.

Further mudholes are provided at the top and bottom of the smokebox tube-plate.

Boiler Fittings. A complete set of fittings of the most substantial and approved pattern is supplied as follows: Two special asbestos packed water gauges arranged to close automatically and instantly in the event of a gauge glass being broken fitted with stout brass guards: pressure gauge: whistle: blow-off cock: double safety valves: special gland-packed steam cocks for injector and water lifter: check valves or clack boxes for pump and injector having special cocks between check valve and boiler whereby the boiler may be completely disconnected when under steam: fusible plug: &c., &c.

Firebox. **The GARRETT CORRUGATED FIREBOX of UNEQUALLED ECONOMY, STEAMING CAPACITY and DURABILITY.**

Tests (Boiler and Engine). The boilers of the **Garrett Compound Steam Motor Tractors** are tested by hydraulic pressure (cold water) to 360 lbs. per square inch for a regular working steam pressure of 180 lbs. per square inch.

Every Garrett Steam Motor Tractor has to undergo a prolonged brake trial at its regular working steam pressure, carrying full load and lasting several hours, when indicator diagrams are taken: no Engine is allowed to leave Leiston Works without having passed through this trial satisfactorily.

In addition to the above tests every Garrett Steam Motor Tractor is also tested with a full load behind it over a variety of roads.

Smokebox. Of extra length to arrest sparks.

Firedoor. Fitted with special notched latch to enable air to be admitted through firedoor when desired so as to prevent the emission of smoke if tractor is stopped suddenly in city traffic.

Blast Cock, of special type arranged to be operated from the footplate, to prevent emission of smoke if Tractor is stopped in city traffic.

Tender. The tender is constructed of extra stout mild steel plates with capacious water tank underneath footplate and coal bunker on the top: all corners are flanged by special hydraulic machinery with large "rounding" curves, instead of sharp corners so as to assist cleaning and prevent the accumulation of mud, dirt, &c., &c., and consequent corrosion.

Front (Belly) Tank. In addition to the tank under footplate in tender, a second water tank of large capacity is fitted under boiler barrel in front of the firebox; the two tanks being connected with a suitable pipe having a stop cock for use when it is desired to disconnect the front tank for any purpose.

Ashpan and Gear Covers are of stout steel plates flanged by special hydraulic machinery with large "rounding" curves instead of sharp corners to prevent the accumulation of dirt, &c., and assist cleaning.

Special Drawbar Attachment, so arranged that the draught from the drawbar is taken direct to the main axle by powerful flat rods, without passing through the tender: the drawbar attachment is provided with a special wedge adjustment.

Engine Motion Work is neatly enclosed by sheet steel guards, brass beaded at the edges.

Spring Seat for Steering is provided.

Lagging. Boiler lagged with approved non-conducting material, neatly covered with sheet steel casing, secured with polished brass bands with straining screws round boiler barrel.

Painting. The Engine is painted, lined out and varnished in best locomotive style.

Outfit. With every Garrett Steam Motor Tractor the following outfit is provided: Waterproof cover of best Willesden or equal cloth: one pair of large head lamps: one tail lamp with sliding shutters and one water gauge lamp for night work: tool box with all necessary spanners, also hammer, caulking tool and oil can: stoking tools: tube brush: complete set of spuds or clips, also frost pegs for hind driving wheels: fusible plug: &c., &c.

Extras. The following, when ordered specially, can be fitted to the Garrett Steam Motor Tractors at extra charge (prices on application):—

- Specially large, Colonial type firebox (of the Garrett Corrugated type, pages 4 and 5).
- Oil-burning apparatus.
- Special "drop" or "rocking" grates.
- Extra wide travelling wheels.
- Awning (1) over tender or (2) over full length of Engine.
- Extended smokebox and dynamo bracket (see page 65).
- Showmen's fittings (see page 65).
- Extra large front (belly) tank.
- Spark arrestors of various types.
- Small jib or other suitable type of cranes.
- Parts to convert into a steam road roller (see pages 22 and 23).
- Dismantling and close packing for shipment.

Appendix 1
Table of dimensions of tractors & rollers c.1909

† *Reading* roller

* From 1920 list

¶ Superheating added 10 sq. ft.

N.B. The CoReading and the 5 ton and 20 ton compound rollers were never made.

Type of engine	Boiler Barrel Length by Diameter	Tubes: Number	Tubes: Diameter	Working Pressure, lb. per sq. in.	Weight Empty (tons)	Weight in Working Order (tons)	Maximum b.h.p.	Coal per Hour per b.h.p. (lb.)	Water per Hour per b.h.p. (lb.)	Engine Speed, r.p.m.	Gear Ratios: Slow	Gear Ratios: Fast	Speed in m.p.h.: Slow	Speed in m.p.h.: Fast
No. 3 tractor	3′ 1⅛″ × 1′ 8½″	19	1¾″	160	3.75	4.25	18	5.00	39	250	1 : 14.8	1 : 7.66	2.41	4.65
No. 4 S.C. tractor	3′ 10 3/16″ × 1′ 10½″	28	1¾″	160	4.85	5.50	25	4.00	32	230	1 : 24.4	1 : 11.90	1.67	3.44
No. 4 C.D. tractor	3′ 10 3/16″ × 1′ 10½″	28	1¾″	180	4.95	5.60	25	3.25	26	230	1 : 24.4	1 : 11.90	1.67	3.44
*No. 4 C.D. tractor s/heater	3′ 10⅞″ × 1′ 10½″	26	1¾″	180	5.15	5.80	25	3.00	26	230	1 : 24.4	1 : 11.90	1.67	3.44
No. 4 C.D. Colonial tractor	3′ 10 3/16″ × 1′ 10½″	26	1¾″	180	5.25	5.90	25	—	26	230	1 : 24.4	1 : 11.90	1.67	3.44
4 ton S.C. roller	3′ 1⅛″ × 1′ 8½″	19	1¾″	160	4.50	4.75	18	5.00	39	250	§	§	1.70	3.10
5 ton S.C. roller	3′ 10 3/16″ × 1′ 10½″	28	1¾″	160	6.00	6.50	25	4.00	32	230	1 : 22.8	1 : 11.10	1.62	3.33
5 ton C.D. roller	3′ 10 3/16″ × 1′ 10½″	28	1¾″	180	6.25	6.45	25	3.25	26	230	1 : 22.8	1 : 11.10	1.62	3.33
6 ton C.D. roller	3′ 10 3/16″ × 1′ 10½″	28	1¾″	180	8.00	8.50	25	3.25	26	230	1 : 24.4	1 : 11.90	1.68	3.44
8 ton C.D. roller	3′ 10 3/16″ × 1′ 10½″	28	1¾″	180	8.00	8.50	25	3.25	26	230	1 : 24.4	1 : 11.90	1.62	3.33
10 ton C.D. light roller	3′ 10 3/16″ × 1′ 10½″	28	1¾″	180	10.00	10.50	25	3.25	26	230	1 : 24.4	1 : 11.90	1.68	3.44
10 ton "Coreading" roller	4′ 1¾″ × 1′ 10⅝″	26	1¾″	160	10.00	10.50	§	§	—	180	§	§	1.70	3.10
6 ton S.C. roller	4′ 1 3/16″ × 1′ 10⅝″	26	1¾″	140	8.00	8.50	17	4.25	34	225	1 : 29.8	1 : 17.94	1.30	2.23
8 ton S.C. roller	4′ 1 3/16″ × 1′ 10⅝″	26	1¾″	140	9.00	9.50	17	4.25	34	225	1 : 29.8	1 : 17.94	1.37	2.29
†10 ton S.C. roller	4′ 1¾″ × 1′ 10⅝″	26	1¾″	140	10.25	10.75	21	4.25	34	225	1 : 29.8	1 : 17.94	1.40	2.34
10 ton S.C. standard roller	4′ 3 3/16″ × 2′ 1½″	34	1¾″	140	12.50	13.25	25	4.00	32	180	1 : 30.4	1 : 15.30	1.15	2.30
12 ton S.C. roller	4′ 3 3/16″ × 2′ 1½″	34	1¾″	140	13.25	14.00	25	4.00	32	180	1 : 30.4	1 : 15.30	1.15	2.30
10 ton C.D. roller	4′ 3 3/16″ × 2′ 1½″	34	1¾″	160	12.75	13.50	21	3.25	26	180	1 : 30.4	1 : 15.30	1.15	2.30
12 ton C.D. roller	4′ 3⅛″ × 2′ 5½″	39	1¾″	160	14.50	15.25	25	3.10	25	180	1 : 30.4	1 : 15.30	1.16	2.31
15 ton C.D. roller	4′ 3⅛″ × 2′ 5½″	39	1¾″	160	16.25	17.00	25	3.10	25	180	1 : 30.4	1 : 15.30	1.21	2.40
18 ton C.D. roller	4′ 6 11/16″ × 2′ 9½″	46	1¾″	160	19.25	20.50	35	3.00	24	180	1 : 27.9	1 : 16.6	1.38	2.32
20 ton C.D. roller	4′ 6 11/16″ × 2′ 9½″	46	1¾″	160	21.25	22.50	35	3.00	24	180	1 : 27.9	1 : 16.6	1.38	2.32

Type of Engine	Diameter of Crankshaft	Diameter of First Countershaft	Diameter of Second Countershaft	Diameter of Main Axle	Tank Capacity: Tender (gallons)	Tank Capacity: Belly (gallons)	Bunker Capacity: Cubic Feet	Bunker Capacity: Cwt.	Cylinder Diameter	Cylinder Stroke	Flywheel Diameter by Width	Hindwheels Diameter by Width	Fore Wheels Diameter by Width	Extreme Width of Engine	Extreme Length of Engine	Height to Top of Flywheel	Grate Area, square feet	Heating Surface, square feet
No. 3 tractor	2½″	2¾″	—	3½″	53	60	4	1.6	5½″	9″	2′ 6″ × 4½″	4′ 0″ × 8″	3′ 0″ × 5″	4′ 6″	12′ 8⅜″	6′ 1″	2.1	42.8
No. 4 S.C. tractor	2¾″	3½″	—	4½″	65	85	9½	3.8	6½″	10″	3′ 3″ × 5″	5′ 0″ × 1′ 0″	3′ 4″ × 6″	5′ 10⅞″	14′ 5″	7′ 3″	3.0	68.8
No. 4 C.D. tractor	2¾″	3½″	—	4½″	65	85	9½	3.8	4½″ & 6¾″	10″	3′ 3″ × 5″	5′ 0″ × 1′ 0″	3′ 4″ × 6″	5′ 10⅞″	14′ 5″	7′ 3″	3.0	68.8
*No. 4 C.D. tractor s/heater	2¾″	3½″	—	4½″	65	75	9½	3.8	4½″ & 6¾″	10″	3′ 3″ × 5″	5′ 0″ × 1′ 0″	3′ 7¾″ × 6″	6′ 0″	14′ 10″	7′ 3″	3.0	¶68.12
No. 4 C.D. Colonial tractor	2¾″	3½″	—	4½″	65	75	9½	3.8	4½″ & 6¾″	10″	3′ 3″ × 5″	5′ 0″ × 1′ 0″	3′ 7¾″ × 6″	6′ 0″	15′ 9″	7′ 3″	4.38	73.14
4 ton S.C. roller	2½″	2¾″	—	3½″	53	—	4	1.6	5½″	9″	2′ 6″ × 4½″	4′ 0″ × 1′ 0″	3′ 0″ × 1′ 8½″	5′ 2″	14′ 3	6′ 1″	2.10	42.80
5 ton S.C. roller	2¾″	3½″	—	4½″	65	—	9½	3.8	6½″	10″	3′ 3″ × 5″	4′ 6″ × 1′ 2″	3′ 3″ × 1′ 10⅞″	5′ 10″	15′ 5¼″	6′ 10″	3.00	68.80
5 ton C.D. roller	2¾″	3½″	—	4½″	65	—	9½	3.8	4¾″ & 6¾″	10″	3′ 3″ × 5″	4′ 6″ × 1′ 2″	3′ 3″ × 2′ 1″	6′ 2″	15′ 5¼″	6′ 10″	3.00	68.80
6 ton C.D. roller	2¾″	3½″	—	4½″	65	—	9½	3.8	4¾″ & 6¾″	10″	3′ 3″ × 5″	5′ 0″ × 1′ 3″	3′ 4½″ × 2′ 1″	6′ 2″	15′ 6″	7′ 0″	3.00	68.80
8 ton C.D. roller	2¾″	3½″	—	4½″	65	—	9½	3.8	4½″ & 6¾″	10″	3′ 3″ × 5″	4′ 10″ × 1′ 3″	3′ 4″ × 2′ 0⅞″	6′ 2¼″	15′ 5⅞″	6′ 11″	3.00	68.80
10 ton C.D. light roller	2¾″	3½″	—	4½″	65	—	9½	3.8	4½″ & 6¾″	10″	3′ 3″ × 5″	5′ 0″ × 1′ 5″	3′ 6″ × 2′ 0⅞″	6′ 6¼″	15′ 6¾″	7′ 0″	3.00	68.80
10 ton "Coreading" roller	2¾″	2¾″	3″	4½″	65	—	4	1.6	5¼″ & 8½″	12″	3′ 3″ × 4″	5′ 3″ × 1′ 5″	3′ 8½″ × 1′ 6 11/16″	6′ 0″	16′ 8⅜″	7′ 3″	4.12	72.04
6 ton S.C. roller	2¾″	2¾″	3″	4″	65	—	4	1.6	6½″	10″	3′ 3″ × 2½″	5′ 0″ × 1′ 3″	3′ 3″ × 1′ 6 11/16″	5′ 5″	15′ 3¾″	7′ 0″	3.40	69.04
8 ton S.C. roller	2¾″	2¾″	3″	4″	65	—	4	1.6	6½″	10″	3′ 3″ × 2½″	5′ 1½″ × 1′ 4″	3′ 4½″ × 1′ 6 11/16″	5′ 7″	15′ 4½″	7′ 1⅜″	3.40	69.04
†10 ton S.C. roller	2¾″	2¾″	3″	4½″	65	—	4	1.6	7″	10″	3′ 3″ × 4″	5′ 3″ × 1′ 5″	3′ 8½″ × 1′ 6 11/16″	6′ 0″	16′ 8⅜″	7′ 3″	4.12	72.04
10 ton S.C. standard roller	3″	3½″	3½″	4⅞″	105	—	9½	4	8″	12″	4′ 2″ × 5″	5′ 6¾″ × 1′ 5″	4′ 0″ × 2′ 0½″	6′ 6⅜″	18′ 9⅜″	8′ 3″	4.40	92.00
12 ton S.C. roller	3″	3½″	3½″	4⅞″	105	—	9½	4	8″	12″	4′ 2″ × 5″	5′ 6¾″ × 1′ 6″	4′ 0″ × 2′ 2½″	6′ 8⅜″	18′ 9⅜″	8′ 3″	4.40	92.00
10 ton C.D. roller	3″	3½″	3½″	4⅞″	105	—	9½	4	5½″ & 8½″	12″	4′ 2″ × 5″	5′ 6¾″ × 1′ 5″	4′ 0″ × 2′ 0½″	6′ 6⅜″	18′ 9⅜″	8′ 3½″	4.40	92.00
12 ton C.D. roller	3½″	3½″	3½″	5½″	127	—	10	4.2	5½″ & 9″	12″	4′ 2″ × 5″	5′ 9″ × 1′ 6″	4′ 0″ × 2′ 2½″	7′ 1⅜″	18′ 8½″	7′ 11⅞″	4.75	103.8
15 ton C.D. roller	3½″	3½″	3½″	5½″	127	—	10	4.2	5½″ & 9″	12″	4′ 2″ × 5″	5′ 9″ × 1′ 8″	4′ 3″ × 2′ 2½″	7′ 4⅞″	18′ 10″	8′ 0⅜″	4.75	103.8
18 ton C.D. roller	3½″	3½″	4″	5½″	185.4	—	13.6	5.6	6″ & 10½″	12″	4′ 6″ × 6″	6′ 0″ × 1′ 8″	4′ 6″ × 2′ 5¼″	7′ 9½″	20′ 5½″	8′ 11½″	6.09	130.6
20 ton C.D. roller	3½″	3½″	4″	5½″	185.4	—	13.6	5.6	6″ & 10¾″	12″	4′ 6″ × 6″	6′ 0″ × 1′ 9″	4′ 6″ × 2′ 5¼″	7′ 11½″	20′ 5½″	8′ 11½″	6.09	130.6

Appendix 2 - Telegraphic Codes

In the years when rapid communication of messages was dependent upon the use of the telegraph it was essential to keep messages terse, for reasons of economy as much as for any other. To this end various telegraphic codes were available. Of these ABC and Bentley's were frequently employed but others used included Engineering, Liebers, Western Union, A1, New General & Mining, and Century Standard. Many companies used their own private codes. Garretts used both the ABC 5th Edition and a private system of their own. This latter required each engine type and every part on the spare parts list for that engine to have a code-name. The code-name for a complete No.4CD tractor was COQUARTRAC and it was accompanied by an illustrated and codified part list in which every sub-assembly was allocated a code-word beginning with 'R' and each individual component a code-name commencing with 'S'. Thus the belly tank complete with all fittings became 'RUODLAND' and, for example, the manhole cover 'SPELLANDO'. By issuing the parts list in English, French, German, and Spanish the problems of translation were much reduced. Garretts' instruction to users was:

> When ordering parts by telegram it is only necessary to give the code-word and the number of the engine for which the part is required.

This system was introduced as part of Alfred Garrett's programme of administrative reform in the period 1903/04. The codes for the various types covered by this book were as follows:

Tractors

No.3 tractor	TERTRAC
No.4 single cylinder tractor	QUARTRAC
No.4 compound tractor	COQUARTRAC

3-Shaft Rollers

4 ton single	MARQUIS
5 ton single	EARL
8 ton compound	PRINCE
10 ton compound	BARON

4-Shaft Rollers

6 ton single	MONARCH	10 ton compound (standard)	AUTOCRAT
8 ton single	RULER	12 ton compound	PRESIDENT
10 ton single (light)	READING	15 ton compound	REGENT
10 ton single (standard)	KING	18 ton compound	CZAR
12 ton single	QUEEN		
15 ton single superheated	WAZIR		

Other superheated rollers had the prefix SUP added to the code for the non-superheated version.
Proposals to build a 5 ton compound roller, based upon the 4CD tractor, with the code-name DUKE; a 20 ton compound roller code-named EMPEROR; and a compound version of the READING (to be code-named COREADING) were never pursued.

Using codes undoubtedly saved money in telegram fees. It was much cheaper, for example, to say AUTOCRAT than '10 ton standard compound 4-shaft roller' but there were disadvantages as well. The code-words for parts were mostly artificially concocted non-words such as SORCIERES, SORCIONE, SORCOLOSA, and SORDASTRO used successively to denote components of the crankshaft brackets, incomprehensible to even a reasonably literate telegraphist. Consequently mistakes of transcription did occur and often clients were advised by makers, though, curiously enough, not by Garretts, to write the draft of their telegram in capitals.

Appendix 3 - Key to Prefixes applied to Drawings & Patterns

The first rollers, based upon the No.6 single cylinder traction engine, used the same prefix - i.e. 'T'. Very soon, however, the prefix was altered to 'ST' as in the list below.

Tractors

TR	No.3 single cylinder tractor
MR	No.4 single cylinder tractor
CR	No.4 compound tractor
DR	No.5 compound tractor
BCR	No.4 superheated compound tractor

Rollers

RL	4 ton single cylinder 3-shaft roller
RM	5 ton single cylinder 3-shaft roller
AS	6 & 8 ton single cylinder 4-shaft rollers
ST	10 & 12 ton single cylinder 4-shaft rollers
CM	8 & 10 ton compound 3-shaft rollers
AU	10 ton compound 4-shaft roller
CT	12 & 15 ton 4-shaft rollers
CU	18 & 20 ton compound 4-shaft rollers
BST	10 & 12 ton superheated single cylinder rollers
BKT	15 ton superheated single cylinder roller

Appendix 4 - Indexes of Garrett Steam Tractors & Rollers

In the two indexes (one for rollers and one for tractors) which follow the following conventions have been observed:

[i] The registration number quoted is that with which the engine left the Works or, failing that, the earliest thought to have been applied to it. Many, however, were re-registered later in their lives.
The late Alan Duke devoted many years of his leisure time to tracing and recording the registration details of steam vehicles but some registration authorities, such as County Down, had discarded the information before he could reach it. Alan and I collaborated in collating Garrett registration numbers but a few eluded our joint efforts.

[ii] In the case of engines sold via agents the ultimate purchaser's name is given where known, but in the absence of this knowledge only the name of the agent is stated.

[iii] For the most part, the dates given for delivery are taken from the Order books, supplemented where there are gaps from this source, by information taken from the delivery note books which recorded the dates when vehicles left the manufacturing process and were taken into stock. There were cases in which engines spent extended periods in stock either because they had been built as speculation or were kept for exhibition at the seasonal succession of shows. The keepers of both sets of books were liable to the occasional error.

[iv] Abbreviations used:

cd = compound	tr = tractor
sc = single cylindered	str = showman's tractor
sh = superheated	conv = convertible
col = colonial firebox	rr = road roller

[v] Only the initial purchaser or agent is given. No attempt has been made to list subsequent owners.

***Fig.148.** A rare full rear view of an unidentified late type 4CD tractor newly completed (the strake rivets are unmarked) outside the Erecting Shop and facing out of the Haylings Road gate of the Town Works, c.1920.*

ROAD ROLLERS

Works No.	Regn. No.	Date	Type	Class	Owner	Notes
21411	Export	02/06/98	10t sc rr		Burn & Co., Howrah.	
21494	Export	24/08/98	10t sc rr		Burn & Co., Howrah.	
22393	Export	03/01/00	10t sc rr		Burn & Co., Howrah.	
23039	Export	03/12/00	10t sc rr		Mandalay Municipality, via Burn	Still at work in 1951.
23147	Export	04/04/01	10t sc rr		W&J Scheid, Limburg a.d. Lahn, Germany.	Without rollers.
23376	Export	04/05/01	10t sc rr		W&J Scheid, Limburg a.d. Lahn, Germany.	Without rollers.
23377	HR3887	10/06/01	10t sc rr		Wm.Rawlings, Collingbourne Ducis, Wilts.	Royal Counties Show."General French".Exported 1952.
23476	Export	01/08/01	10t sc rr		Cawnpore Municipality, via Burn.	
23486	Export	19/09/01	10/12t sc rr		A.Henninger & Co. Darmstadt, Germany	
23501		29/03/02	10t sc rr		Wm.Rawlings, Collingbourne Ducis, Wilts.	"Lord Kitchener"
23546	Export	14/03/02	10t sc rr		W&J Scheid, Limburg a.d. Lahn, Germany.	Without rollers.
23547	Export	23/05/02	10t sc rr		Burn & Co., Howrah.	
23607	Export	31/01/02	15t cd rr		A.Henninger & Co. Darmstadt, Germany	Tapered rollers.
23633	Export	14/02/03	15t cd rr		W&J Scheid, Limburg a.d. Lahn, Germany.	Without rollers, flywheel etc.
23642		01/10/02	10t sc rr		G.E.Gould, Ipswich, Suffolk.	Suffolk Show June '02
23677	Export	05/05/02	18t cd rr		A.Henninger & Co. Darmstadt, Germany	German Royal Show, Mannheim.
23695	Export	26/06/02	15t cd rr		Th.Ohl, Limburg a.d. Lahn, Germany	Without rollers, axle etc.
23729	Export	29/04/03	18t cd rr		B.Ruthemeyer, Soest, Germany.	Without rollers, etc.
23745	HR3583	30/06/02	10t sc rr		Wm.Rawlings, Collingbourne Ducis, Wilts.	Royal Counties Show. "Highwayman". Exported 1952
23889	Export	19/09/02	10t sc rr		Burn & Co., Howrah.	
23952	BJ5964	01/12/02	10t sc con		G.R.LeGrys, Heveningham, Suffolk.	Scarifier and awning fitted.
23964		06/10/02	10t sc rr		A.Bailey, Wymondham, Norfolk.	
23993	Export	14/02/03	8t sc rr		W&J Scheid, Limburg a.d. Lahn, Germany.	Without rollers.
24013	HR4390	19/02/03	10t sc rr	"Reading"	H.Lewis & Sons, Reading, Berks.	
24014		01/06/03	10t sc rr	"Reading"	RG&S, Leiston, Suffolk.	Suffolk Show June '03. Rebuilt as 24805
24015	Export	24/03/04	9t sc rr	"Reading"	A.Henninger & Co. Darmstadt, Germany	
24233	Export	16/01/03	10t sc rr		Burn & Co., Howrah.	
24285	Export	14/05/03	10t sc rr		Th.Ohl, Limburg a.d. Lahn, Germany	Without rollers, etc.
24286	Export	14/05/03	15t cd rr		Th.Ohl, Limburg a.d. Lahn, Germany	Without rollers, etc.
24288	Export	20/02/03	10t sc rr		Th.Ohl, Limburg a.d. Lahn, Germany	Without rollers, flywheel etc.
24319	Export	10/03/03	10t sc rr		A.Henninger & Co. Darmstadt, Germany	Without front rollers and hind roller rims.
24347	Export	30/03/03	10t sc rr		A.Henninger & Co. Darmstadt, Germany	Complete
24353	Export	09/06/03	18t cd rr		B.Ruthemeyer, Soest, Germany.	Without rollers, etc.
24400	Export	08/07/03	10t sc rr		Osias Ausschnitt, Galatz, Roumania.	Devon County Show May '03, Royal June '03
24426	Export	23/05/03	8t sc rr		Basalt Maatschappij, Rotterdam, via H.F.Roll.	
24434	Export	23/04/04	15t cd rr		W&J Scheid, Limburg a.d. Lahn, Germany.	Without rollers, etc.
24468		21/08/03	15t cd rr		Belfast Corporation.	No. 4
24526	YA1767	07/10/03	10t sc rr		W.R.Johnson, E.Dereham, Norfolk.	
24556	Export	06/10/03	10t sc rr		Mysore Corporation, via Burn.	
24565	Export	16/10/03	8t sc rr	"Ruler"	Th.Ohl, Limburg a.d. Lahn, Germany	Without rollers, etc.
24725	Export	29/02/04	8t sc rr	"Ruler"	Port Swettenham, via Burn.	
24726	Export	29/02/04	6t sc rr	"Monarch"	District Board Engineer, Godavari, via Burn.	
24804	Export	11/04/04	18t cd rr	"Czar"	W&J Scheid, Limburg a.d. Lahn, Germany.	Without rollers, etc.
24805	Export	08/03/04	9t sc rr	"Reading"	W&J Scheid, Limburg a.d. Lahn, Germany.	Originally 24014. Without rollers etc.
24911	Export	30/03/04	10t sc rr	"King"	W&J Scheid, Limburg a.d. Lahn, Germany.	Without rollers, etc.
24990	Export	15/08/04	6t sc rr	"Monarch"	Burn & Co., Howrah.	Spares sent 1939 to Jammu Irrigation Division, Kashmir State..
24993	Export	20/06/04	10t sc rr	"King"	A.Henninger & Co. Darmstadt, Germany	Complete
25013	Export	31/08/04	6t sc rr	"Monarch"	Soc. Anglo-Espanola de Motores & Co., Madrid, Spain.	For Public Works of Spain.
25020	Export	28/09/04	8t sc rr	"Ruler"	Burn & Co., Howrah.	
25028	Export	31/08/04	10t sc rr	"King"	Soc.Anglo-Espanola de Motores & Co., Madrid, Spain.	For Public Works of Spain.
25053	Export	17/10/04	10t sc rr	"King"	Calcutta Corporation, India, via Burn.	
25054	Export	08/10/04	10t sc rr	"King"	Mysore Corporation, via Burn.	Scarifier fitted.
25143	Export	06/12/04	10t sc rr	"King"	Soc.Anglo-Espanola de Motores & Co., Madrid, Spain.	
25256	Export	24/01/05	10t sc rr	"King"	Cawnpore Municipality,via Burn.	
25282	Export	18/01/05	8t sc rr	"Ruler"	Burn & Co., Howrah.	
25307	Export	08/02/05	12t sc rr	"Queen"	Province of Palermo, via Grimaldi.	With steel brackets to carry extra weights awning etc.
25327	Export	18/03/05	15t cd rr	"Regent"	Soc. Anglo-Espanola de Motores & Co., Madrid, Spain.	
25346	Export	16/05/05	6t sc rr	"Monarch"	Thos.Wilson & Co., London	For Bombay.
25354	Export	19/04/05	10t sc rr	"King"	Th.Ohl, Limburg a.d. Lahn, Germany	Without rollers, etc.
25374	Export	03/05/05	8t sc rr	"Ruler"	Palmer Graham, Rostoff on Don, Russia.	
25378	Export	05/05/05	10t sc rr	"King"	A.Henninger & Co. Darmstadt, Germany	Complete
25380	Export	29/05/05	15t cd rr	"Regent"	Government of Mendoza, Argentina, via Mattos.	
25384		11/07/05	15t cd rr	"Regent"	Belfast Corporation.	No. 5
25387	Export	13/05/05	10t sc rr	"King"	W&J Scheid, Limburg a.d. Lahn, Germany.	Without rollers, etc.
25388	Export	22/05/05	10t sc rr	"King"	A.Henninger & Co. Darmstadt, Germany	Cast steel rims to hind rollers.
25402	Export	04/07/05	10t sc rr	"King"	Th.Ohl, Limburg a.d. Lahn, Germany	Without rollers, etc.
25424	Export	02/08/05	10t cd rr	"Autocrat"	Palermo, Sicily, via Grimaldi.	
25442	Export	29/06/05	6t sc rr	"Monarch"	A.Henninger & Co. Darmstadt, Germany	
25462	Export	21/07/05	15t cd rr	"Regent"	Th.Ohl, Limburg a.d. Lahn, Germany	Without rollers, etc.
25465	Export	15/08/05	10t sc rr	"King"	Burn & Co., Howrah.	
25507	Export	25/09/05	12t sc con	"Queen"	Provincia de Mendoza, Argentina, via Mattos.	
25528	Export	01/09/05	8t sc rr	Special	Trapani, Sicily, via Grimaldi.	
25535	Export	02/10/05	8t sc rr	"Ruler"	Ministerio Obras Publicas, B.A. via Mattos.	
25558	Export	30/11/05	6t sc rr	"Monarch"	Rangoon, via Burn.	
25571	Export	14/12/05	8t sc rr	"Ruler"	Th.Ohl, Limburg a.d. Lahn, Germany	Without rollers, fork, etc.
25647	Export	06/12/05	10t sc rr	"King"	Sociedad Puerto del Rosario, via Mattos	
25649	Export	10/01/06	6t sc rr	"Monarch"	Burn & Co., Howrah.	Calcutta
25713	Export	21/02/06	10t sc rr	"King"	W&J Scheid, Limburg a.d. Lahn, Germany.	Without rollers, etc.
25714	Export	21/02/06	10t sc rr	"King"	W&J Scheid, Limburg a.d. Lahn, Germany.	Without rollers, etc.
25728	Export	15/03/06	10t sc rr	"King"	City of Rome, via Grimaldi.	Milan Exhibition.
25755	Export	04/04/06	9t sc rr	"Ruler"	A.Henninger & Co. Darmstadt, Germany	
25757	Export	13/02/06	8t sc rr	"Ruler"	Italian State Railways, via Grimaldi.	
25758	Export	27/03/06	12t cd rr	"President"	A.P.Muscate GmbH, Danzig & Dirschau.	
25777	Export	11/04/06	10t sc rr	"King"	Burn & Co., Howrah.	
25785	Export	16/03/06	10t sc rr	"King"	W&J Scheid, Limburg a.d. Lahn, Germany.	Without rollers, etc.
25789	Export	14/04/06	10t cd rr	"Autocrat"	Nickel & Seitz, Mannheim, Germany, via Henninger.	Without rollers, etc.
25818	Export	11/05/06	14t cd rr	Special	Local Works of Guipuzcoa, Spain, via Harker Sumner.	
25837	Export	17/04/06	6t sc rr	"Monarch"	Strassenwalzen-Betrieb , Niederlahnstein, Germany.	Without rollers, etc.
25862	Export	01/05/06	10t sc rr	"King"	W&J Scheid, Limburg a.d. Lahn, Germany.	Without rollers, fork, etc.
25868	Export	10/05/06	10t sc rr	"King"	Th.Ohl, Limburg a.d. Lahn, Germany	Without rollers, fork, etc.
25869	Export	02/06/06	12t cd rr	"President"	Strassenwalzen-Betrieb , Niederlahnstein, Germany.	Without rollers.
25870	Export	03/07/06	12t cd rr	"President"	Strassenwalzen-Betrieb , Niederlahnstein, Germany.	Without rollers.
25912	Export	27/06/06	6t sc rr	"Monarch"	Ministerio Obras Publicas, B.A. via Mattos.	
25931	Export	06/07/06	8t sc rr	"Ruler"	Th.Ohl, Limburg a.d. Lahn, Germany	Without rollers, etc.
25956	Export	29/08/06	6t sc rr	"Monarch"	Burn & Co., Howrah.	
25983	Export	22/12/06	10t sc rr	"King"	Th.Ohl, Limburg a.d. Lahn, Germany	Without rollers, etc.

Works No.	Regn. No.	Date	Type	Class	Owner	Notes
25995	Export	22/12/06	10t sc rr	"King"	Th.Ohl, Limburg a.d. Lahn, Germany	Without rollers, etc.
26007	Export	20/10/06	15t cd rr	"Regent"	Ministerio Obras Publicas, B.A.via Mattos.	
26010	Export	22/09/06	8t sc rr	"Ruler"	Government of Cordoba, via Mattos.	
26011	Export	20/10/06	15t cd con	"Regent"	Government of Cordoba, via Mattos.	
26022	Export	06/11/06	8t sc rr	"Ruler"	Burn & Co., Howrah.	
26036	Export	28/09/06	10t sc rr	Special	Th.Ohl, Limburg a.d. Lahn, Germany	Without rollers, etc.
26072	Export	15/11/06	12t cd rr	"President"	Province of Catania, Sicily, via Grimaldi.	
26140	Export	21/01/07	10t cd rr	"Autocrat"	Invercargill Borough Council, NZ, via Macalister	
26310	Export	11/02/07	12t cd rr	"President"	A.P.Muscate GmbH, Danzig & Dirschau.	Without rollers.
26311	Export	26/02/07	12t cd rr	"President"	A.P.Muscate GmbH, Danzig & Dirschau.	Without rollers.
26353	Export	06/02/07	10t sc rr	"King"	Rostoff Town Council, Russia, via Palmer Graham	
26481	Export	30/03/07	10t sc rr	"King"	A.Henninger & Co. Darmstadt, Germany	Front rolls complete: rears, spiders only
26485	Export	27/02/07	6t sc rr	"Monarch"	Burn & Co., Howrah.	
26486	Export	23/04/07	8t sc rr	"Ruler"	District Engineer of Gaya, India, via Burn.	
26487	Export	09/07/07	10t sc rr	"King"	Burn & Co., Howrah.	
26509	Export	31/05/07	12t cd rr	"President"	City of Maastricht, via H.F.Roll.	Renewable rims all round
26527	Export	13/07/07	12t cd rr	"President"	Grimaldi, Italy.	3" thick rolls; speed indicator.
26531	Export	13/08/07	8t sc rr	"Ruler"	Bassein Municipality, Burma via Burn	
26556	Export	10/06/07	8t sc rr	"Ruler"	H. Boesewinkel, Zutphen, via H.F.Roll	
26567	Export	04/07/07	6t sc rr	"Monarch"	Strassenwalzen-Betrieb , Niederlahnstein, Germany.	Without rollers.
26586	Export	26/07/07	12t cd rr	"President"	Public Works of Cuba, via Czarnikow.	
26587	Export	26/07/07	12t cd rr	"President"	Public Works of Cuba, via Czarnikow.	
26588	Export	29/08/07	12t cd rr	"President"	Public Works of Cuba, via Czarnikow.	
26589	Export	29/08/07	12t cd rr	"President"	Public Works of Cuba, via Czarnikow.	
26590	Export	29/08/07	12t cd rr	"President"	Public Works of Cuba, via Czarnikow.	
26613	Export	21/12/07	15t cd rr	"Regent"	Th.Ohl, Limburg a.d. Lahn, Germany.	Without rollers, etc.
26617	Export	09/10/07	10t cd rr	"Autocrat"	Ministerio Obras Publicas, B.A.via Mattos.	
26624	Export	20/09/07	10t sc rr	"King"	Italian State Railways, via Grimaldi.	
26625	Export	11/10/07	10t sc rr	"King"	Italian State Railways, via Grimaldi.	
26626	Export	24/09/07	10t sc rr	"King"	Italian State Railways, via Grimaldi.	
26627	Export	11/11/07	12t cd rr	"President"	Ministerio Obras Publicas, B.A.via Mattos.	
26628	Export	28/10/07	12t cd rr	"President"	Ministerio Obras Publicas, B.A.via Mattos.	
26686	Export	05/11/07	6t sc rr	"Monarch"	Burn & Co., Howrah.	Working for Tata Iron & Steel Co. in 1939
26723	Export	19/11/07	8t sc rr	"Ruler"	Burn & Co., Howrah.	
26731	BJ8490	21/11/07	10t sc rr	"King"	S.H.Carter, Wrentham, Suffolk.	
26735	Export	06/01/08	6t sc rr	"Monarch"	Burn & Co., Howrah.	Special rollers and forks
26755	Export	24/12/07	10t cd rr	"Autocrat"	Spanish Govt, Tarragona, via S.A.E.M.	
26756	Export	04/01/08	10t cd rr	"Autocrat"	Spanish Govt. Alicante, via S.A.E.M.	
26757	Export	26/02/08	10t cd rr	"Autocrat"	Spanish Govt. Bilbao,via S.A.E.M.	
26758	Export	26/02/08	10t cd rr	"Autocrat"	Spanish Govt. Bilbao,via S.A.E.M.	
26759	Export	11/02/08	10t cd rr	"Autocrat"	Spanish Govt. Cadiz,via S.A.E.M.	
26838	Export	13/03/08	6t sc rr	"Monarch"	Strassenwalzen-Betrieb, Niederlahnstein, Germany.	Without rollers.
26839	Export	05/03/08	6t sc rr	"Monarch"	Strassenwalzen-Betrieb, Niederlahnstein, Germany.	Without rollers.
27020	Export	31/01/08	6t sc rr	"Monarch"	Valparaizo, Chile.	via Schill, Seebohm & Co. Manchester.
27024	AF4435	25/01/08	10t sc rr	"King"	R.Dingle & Sons, Stoke Climsland, Cornwall.	"St.Austell"
27040	Export	12/03/08	8t sc rr	"Ruler"	Italian Govt, Palermo, Sicily via Grimaldi	
27041	Export	12/03/08	12t cd rr	"President"	Italian Govt, Palermo, Sicily via Grimaldi.	
27047	Export	18/02/08	8t sc rr	"Ruler"	A.Henninger & Co. Darmstadt, Germany	Without roller rims.
27048	Export	02/04/08	6t sc rr	"Monarch"	A.Henninger & Co. Darmstadt, Germany	
27068	Export	12/03/08	10t sc rr	"King"	Italian State Rlys, Naples, Italy via Grimaldi.	
27100	Export	18/04/08	10t cd rr	"Autocrat"	Barcelona Municipality, Spain.	via Whitehead, Sumner, Harker. Bomford scarifier
27145	Export	01/05/08	8t sc rr	"Ruler"	H. Boesewinkel, Zutphen, via H.F.Roll.	
27152	Export	20/05/08	8t sc rr	"Ruler"	Philipp, Dresden, via Henninger.	Without roller rims.
27154	Export	25/04/08	10t sc rr	"King"	Gebr.Bobe, Dresden, via Henninger.	Without rollers.
27155	Export	06/06/08	8t sc rr	"Ruler"	Th.Ohl, Limburg a.d. Lahn, Germany	Without rollers, etc.
27173	Export	21/05/08	12t cd rr	"President"	Ministerio Obras Publicas, B.A.via Mattos	
27174	Export	31/07/08	12t cd rr	"President"	Ministerio Obras Publicas, B.A. via Mattos.	
27175	Export	11/08/08	10t cd rr	"Autocrat"	Ministerio Obras Publicas, B.A.via Mattos	
27205	Export	07/07/08	10t sc rr	"King"	H. Boesewinkel, Zutphen, via H.F.Roll	
27224	Export	11/06/08	18t cd rr	"Czar"	A.Henninger & Co. Darmstadt, Germany	Without rollers.
27251	Export	13/10/08	8t sc rr	"Ruler"	Burn & Co., Howrah.	
27252	Export	13/10/08	10t sc rr	"King"	Burn & Co., Howrah.	
27256	Export	01/09/08	6t sc rr	"Monarch"	Burn & Co., Howrah.	
27276	Export	29/09/08	6t sc rr	"Monarch"	Municipal Council Cocanada, via Burn.	
27277	Export	16/10/08	12t cd rr	"President"	City of Genoa, Italy, via Grimaldi.	
27278	Export	19/11/08	10t cd lt rr	"Baron"	City of Genoa, Italy, via Grimaldi.	
27323	Export	14/10/08	10t cd rr	"Autocrat"	H.W.Jaartsveld te Lichtenvoorde via H.F.Roll	No.1
27347	Export	30/12/08	8t sc rr	"Ruler"	Dutch Govt. of Soerabaia, via H.F.Roll	
27475	AF4436	16/12/08	10t sc rr	"King"	R.Dingle & Sons, Stoke Climsland, Cornwall.	"Dartmoor"
27564	Export	22/02/09	10t cd rr	"Autocrat"	Nickel & Seitz, Mannheim, via Henninger.	C.s.fork, without roller rims.
27578	Export	10/03/09	10t cd lt rr	"Baron"	Egyptian Government, via T.Cook.	
27625	Export	26/02/09	10t sc rr	"King"	H. Boesewinkel, Zutphen, via H.F.Roll	No.8
27628	Export	15/02/09	6t sc rr	"Monarch"	Burn & Co., Howrah.	
27632	Export	02/03/09	14t sc rr	"Queen"	J.De Groot, Lierre, Belgium, via Weber.	"Gaston"
27677	Export	03/04/09	14t sc rr	"Queen"	Fl.Petit, Entrepreneur, Anvers, via Weber	"Annie"
27709	Export	13/04/09	6t sc rr	"Monarch"	Strassenwalzen-Betrieb , Niederlahnstein, Germany.	Without rollers.
27710	Export	13/03/09	6t sc rr	"Monarch"	Strassenwalzen-Betrieb , Niederlahnstein, Germany.	Without rollers.
27724	Export	15/06/09	10t sc col	"King"	Burn & Co., Howrah.	Shipped to Bombay
27754	Export	16/06/09	10t cd rr	"Autocrat"	Ufficio Technico Provinciale, Catania, Sicily.	via Grimaldi.
27782	Export	26/02/09	8t sc rr	"Ruler"	W&J Scheid, Limburg a.d. Lahn, Germany.	Without rollers.
27790	Export	12/06/09	8t sc rr	"Ruler"	Maatschapij tot Exploitatie van Stoomwalzen.	Lichtenvoorde. via H.F.Roll
27839	Export	09/07/09	10t cd lt rr	"Baron"	City of Zurich, Switzerland, via von Arx.	
27857	Export	14/07/09	8t sc rr	"Ruler"	Th.Ohl, Limburg a.d. Lahn, Germany	Without rollers, etc.
27870	Export	04/08/09	10t cd rr	"Autocrat"	M.E.Mattos, Buenos Aires, Argentina.	
27871	Export	24/08/09	10t cd rr	"Autocrat"	M.E.Mattos, Buenos Aires, Argentina.	
27889	Export	13/08/09	10t sc rr	"King"	Xavier Hayen, Zepperen, via Weber	"Virginie"
27892	Export	03/09/09	8t sc rr	"Ruler"	Th.Ohl, Limburg a.d. Lahn, Germany	Without rollers, etc.
27902	Export	18/08/09	12t cd rr	"President"	Ayuntamiento de Nereva Paz, Cuba,via Czarnikow	
27925		24/08/09	10t sc rr	"King"	H.Cummins, Coleraine,	
27934	Export	04/09/09	8t sc rr	"Ruler"	H.Boesewinkel, Zutphen,via H.F.Roll	No. 9
27953	Export	30/11/09	10t cd rr	"Autocrat"	City of Copenhagen, Denmark, via V.Andersen.	Damptromle No. 4
27955	Export	20/10/09	12t cd rr	"President"	Italian State Railways, via Grimaldi.	
27972	Export	15/10/09	12t cd rr	"President"	Messina, Sicily, via Grimaldi.	
27975	Export	10/12/09	12t cd rr	"President"	Verona, Italy, via Grimaldi.	
27993	Export	09/11/09	6t sc rr	"Monarch"	Crown Mine Ltd., Johannesburg, SA	
28046	Export	17/12/09	10t cd rr	"Autocrat"	Grimaldi, Italy.	
28048	Export	22/12/09	5t sc rr	"Earl"	Th.Ohl, Limburg a.d. Lahn, Germany	Without rollers.
28049	Export	16/07/10	10t sc sh rr	"Supking"	Grimaldi, Milan, Italy.	

Works No.	Regn. No.	Date	Type	Class	Owner	Notes
28050	Export	24/11/09	6t sc rr	"Monarch"	Burn & Co., Howrah.	New boiler 1931
28069	Export	09/03/10	10t cd sh lt	"Supbaron"	M.E.Mattos, Buenos Aires, Argentina.	Exposicion Centenario.
28070	Export	24/12/09	6t sc rr	"Monarch"	A.Henninger & Co. Darmstadt, Germany	Without rollers.
28077	Export	02/04/10	10t cd sh lt	"Supbaron"	Brussels Exhibition via E.Weber	In stock 31/12/10. 29/04/11 to Grimaldi, Turin Exhibition.
28260	AF4437	29/01/10	10t sc rr	"King"	R.Dingle & Sons, Stoke Climsland, Cornwall.	"Cornishman"
28273	Export	26/03/10	6t sc rr	"Monarch"	A.Henninger & Co. Darmstadt, Germany	Without roller rims.
28274	Export	18/04/10	10t sc rr	"Reading"	A.Henninger & Co. Darmstadt, Germany	Without rollers.
28286	Export	29/04/10	6t sc rr	"Monarch"	A.Henninger & Co. Darmstadt, Germany	Without roller rims.
28371	Export	09/04/10	12t cd rr	"President"	Ing.Arrigoni & C., Milan, Italy, via Grimaldi.	Cast steel forks.
28372	Export	09/04/10	10t cd con	"Baron"	Ing.Arrigoni & C., Milan, Italy via Grimaldi	
28374	Export	06/04/10	6t sc rr	"Monarch"	M. E. Stoomwalzen, Lichtenvoorde,via H.F.Roll	
28395	Export	23/05/10	12t cd con	"President"	Gobierno de Cordoba, B.A. via Mattos.	
28399	Export	24/06/10	12t cd con	"President"	Gobierno de Cordoba, B.A.via Mattos	
28401	Export	14/06/10	6t sc rr	"Monarch"	Municipal Council, Cocanada, via Burn.	
28403	Export	15/07/10	15t cd rr	"Regent"	Grimaldi, Milan, Italy.	Cast steel forks.
28408	AO6201	03/08/10	10t cd lt rr	"Baron"	Corporation of Southend, Essex.	"Elizabeth"
28413	Export	03/09/10	9t sc rr	"Reading"	W&J Scheid, Limburg a.d. Lahn, Germany.	Without rollers.
28414	Export	15/08/10	5t sc rr	"Earl"	A.Henninger & Co. Darmstadt, Germany	Cast steel saddle. Without roller rims.
28415	Export	12/09/10	12t cd con	"President"	M.E.Mattos, Buenos Aires, Argentina.	
28417	Export	31/08/10	9t sc rr	"Ruler"	Burn & Co., Howrah.	
28418	Export	31/08/10	9t sc rr	"Ruler"	Burn & Co., Howrah.	
28741	Export	09/02/11	10t cd lt rr	"Baron"	City of Zurich, Switzerland, via von Arx.	
28746	Export	05/10/10	6t sc rr	"Monarch"	Grimaldi, Milan, Italy.	
28764	Export	31/12/10	8t sc rr	"Ruler"	A.Henninger & Co. Darmstadt, Germany	Without roller rims.
28769	Export	07/03/11	12t cd rr	Special	City of Naples, Italy, via Grimaldi.	Special rollers: based on 10 ton roller.
28770	Export	07/12/10	10t cd rr	"Autocrat"	Municipality of Boksburg, via A.Koppel Ltd. Jo'burg.	
28775	Export	14/02/11	9t sc rr	"Reading"	Burn & Co., Howrah.	
28776	Export	17/01/11	6t sc rr	"Monarch"	Burn & Co., Howrah.	
28988	Export	21/03/11	5t sc rr	"Earl"	A.Henninger & Co. Darmstadt, Germany	Without roller rims.
29007	Export	12/04/11	10t cd lt rr	"Baron"	A.P.Muscate GmbH, Danzig & Dirschau.	Without rollers.
29008	Export	18/05/11	8t cd sh lt r	"Supbaron"	Davos Town Council, Switzerland, via von Arx.	
29011	Export	28/04/11	10t cd lt rr	"Baron"	City of St. Moritz, Switzerland, via von Arx.	
29014	Export	16/06/11	8t sc rr	"Ruler"	A.Henninger & Co. Darmstadt, Germany	Without roller rims.
29016	Export	06/04/11	10t sc rr	"King"	A.Henninger & Co. Darmstadt, Germany	Without roller rims.
29018	Export	31/05/11	8t sc rr	"Ruler"	A.Henninger & Co. Darmstadt, Germany	Without roller rims.
29021	Export	06/05/11	6t sc rr	"Monarch"	Strassenwalzen-Betrieb , Niederlahnstein, Germany.	Without rollers.
29022	Export	03/06/11	6t sc rr	"Monarch"	Strassenwalzen-Betrieb , Niederlahnstein, Germany.	Without rollers.
29031	NO3953	10/05/11	10t sc rr	"King"	Z.Fairclough, Clacton on Sea, Essex.	"Clacton Belle". Bomford scarifier.
29760	Export	18/08/11	15t cd rr	"Regent"	Grimaldi, Milan, Italy.	
29762	Export	28/07/11	9t sc rr	"Reading"	Burn & Co., Howrah.	
29763	Export	20/07/11	10t cd sh lt	"Supbaron"	City of Berne, Switzerland, via von Arx.	
29769	Export	29/08/11	6t sc rr	"Monarch"	Burn & Co., Howrah.	
29776	Export	11/10/11	8t sc rr	"Ruler"	Burn & Co., Howrah.	/39 Spares to Kashmir State (J.V.Road Divn.)
30043	Export	15/04/12	8t sc rr	"Ruler"	A.Henninger & Co. Darmstadt, Germany	Without roller rims.
30059	Export	21/03/12	5t sc rr col	"Earl"	H.F.Roll	GOUV:GOUD TINW 1912/1913 MUNTOK on roller.
30060	Export	21/03/12	5t sc rr col	"Earl"	H.F.Roll	GOUV:GOUD TINW 1912/1913 SOENGEILIAT on roller.
30074	AF4438	10/02/12	10t sc rr	"King"	R.Dingle & Sons, Stoke Climsland, Cornwall.	"King George". Allen scarifier.
30095	Export	19/04/12	8t sc rr	"Ruler"	Kotah State P.W.D., Rajputana, via Burn.	
30104	Export	30/12/11	10t cd lt rr	"Baron"	Cretan Government via J.S.Marcantonakis,La Canée	
30105	Export	30/12/11	10t cd lt rr	"Baron"	Cretan Government via J.S.Marcantonakis,La Canée	
30106	Export	09/01/12	10t cd lt rr	"Baron"	Cretan Government via J.S.Marcantonakis,La Canée	
30107	Export	09/01/12	10t cd lt rr	"Baron"	Cretan Government via J.S.Marcantonakis,La Canée	
30108	Export	30/12/11	10t cd lt rr	"Baron"	Cretan Government via J.S.Marcantonakis,La Canée	
30161	Export	27/12/11	15t cd rr	"Regent"	Grimaldi, Naples, Italy.	
30216	Export	27/02/12	10t sc rr	"Reading"	Burn & Co., Howrah.	
30248	Export	12/02/12	16t cd rr	"Regent"	Gobo de Mendoza, B.A.	
30261	Export	23/03/12	12½t cd rr	"Autocrat"	Petone Boro' Council, Wellington, NZ. via Chambers	Special cast rims. 200 p.s.i. instead of 160 p.s.i.
30268	Export	01/06/12	4t sc rr	"Marquis"	Ministerio Obras Publicas, B.A. via Mattos	Rollers to hold water and ribbed rims.
30301	Export	23/03/12	4t sc rr col	"Marquis"	E.Weber, Brussels	For Belgian Congo. No.1.
30302	Export	23/03/12	4t sc rr col	"Marquis"	E.Weber, Brussels	For Belgian Congo. No.2.
30313	Export	22/03/12	12t cd rr	"President"	Scraeyen Frères, Limbourg, via Weber	"Julienne"
30327	Export	16/05/12	15t sc sh rr	"Wazir"	T.Cook & Son (Egypt) Ltd.	
30349	Export	19/04/12	8t cd rr	"Prince"	Ministerio Obras Publicas, B.A.via Mattos	
30357	Export	01/06/12	12t cd rr	"President"	Grimaldi, Turin, Italy.	
30412	Export	23/07/12	10t cd lt rr	"Baron"	Cawnpore Municipality, via Burn	
30851	Export	14/08/12	8t cd rr	"Prince"	J.Chambers & Son Ltd., Auckland, NZ.	Special rollers.
30857	Export	16/07/12	10t sc rr	"King"	Public Works Dept., Neuchatel, Switzerland.	via von Arx. Allen scarifier.
30872	Export	27/07/12	10t cd rr	"Autocrat"	Maribo Amto Vejvaesen, via V.Andersen.	
30907	Export	07/08/12	10t sc rr	"King"	Czarnikow & Co., New York.	
30926	Export	19/09/12	10t cd lt rr	"Baron"	E.H.Hunter & Co.	Shipped to Korea.
30927	YA972	17/07/12	10t sc sh rr	"Supking"	Bridgwater Town Council, Somerset.	Van Putten Water Sprayer.
30928	Export	25/09/12	10t sc rr	"Reading"	Burn & Co., Howrah.	
30990	Export	03/08/12	10t sc sh rr	"Supking"	A.Henninger & Co. Darmstadt, Germany	Without roller rims.
30992	Export	04/11/12	5t sc rr	"Earl"	Burn & Co., Howrah.	Special cast iron rims
30993	Export	22/08/12	8t sc rr	"Monarch"	Burn & Co., Howrah.	
30994	Export	02/09/12	10t cd lt rr	"Baron"	A.Henninger & Co. Darmstadt, Germany	Cast steel rollers.
31000	Export	26/11/12	10t sc rr	"King"	Burn & Co., Howrah.	
31017	Export	20/12/12	8t sc rr	"Ruler"	A.Henninger & Co. Darmstadt, Germany	Without roller rims.
31031	Export	05/10/12	10t cd rr	"Autocrat"	W.Sonesson & Co.	To Hull for export.
31075	Export	06/11/12	12t cd rr	"President"	N.Roba, Bilstain-Dolhain, via Weber.	"Elise"
31094	Export	16/12/12	10t cd lt rr	"Baron"	Burn & Co., Howrah.	Sold as 6 ton
31095	Export	16/12/12	10t cd lt rr	"Baron"	Burn & Co., Howrah.	
31148	Export	26/11/12	5t sc rr	"Earl"	Burn & Co., Howrah.	
31158	Export	31/12/12	8t cd conv	"Prince"	Orenstein & Koppel Ltd.	
31160	Export	31/01/13	4t sc rr	"Marquis"	Burn & Co., Howrah.	
31178	Export	17/12/12	8t cd rr	"Prince"	Grimaldi, Turin, Italy.	
31179	Export	28/01/13	12½t cd rr	"President"	Grimaldi, Turin, Italy.	
31182	Export	17/10/12	8t sc rr	"Ruler"	M.E. van Stoomwalzen, Lichtenvoorde, via H.F.Roll	
31267	Export	31/12/13	5t sc rr	"Earl"	Burn & Co., Howrah.	
31284	Export	11/03/13	6t sc rr	"Earl"	Burn & Co., Howrah.	
31285	Export	18/02/13	8t sc rr	"Monarch"	Burn & Co., Howrah.	
31295	AF4439	31/12/12	12t sc rr	"King"	R.Dingle & Sons, Stoke Climsland, Cornwall.	"One and All"
31317	AF4440	25/01/13	12t sc rr	"King"	R.Dingle & Sons, Stoke Climsland, Cornwall.	"Princess"
31349	Export	17/03/13	10t sc rr	"Reading"	Burn & Co., Howrah.	
31366	Export	09/04/13	12½t sc rr	"King"	Gemeentewerken No.1 Ede, via H.F.Roll.	
31374	Export	16/05/13	12t cd rr	"President"	Grimaldi, Italy.	
31375	Export	26/04/13	12t cd rr	"President"	Grimaldi, Italy.	
31434	Export	14/03/13	6t sc rr	"Monarch"	A.Henninger & Co. Darmstadt, Germany	Without roller rims.

Works No.	Regn. No.	Date	Type	Class	Owner	Notes
31436	Export	01/04/13	10t sc rr	"Reading"	A.Henninger & Co. Darmstadt, Germany	Without roller rims.
31463	Export	10/06/13	10t sc rr	"King"	Etat de Neuchatel, Switzerland, via von Arx.	No.4. /52 Enquiry for rims S.A.Joly Bureau Technique Machines, Lausanne.
31507	Export	11/11/13	8t sc rr	"Monarch"	Burn & Co., Howrah.	
31536	Export	21/07/13	10t sc rr	"Reading"	Burn & Co., Howrah.	
31538	Export	31/12/13	10t cd rr	"Autocrat"	Burn & Co., Howrah.	
31557	Export	19/05/13	10t cd lt rr	"Baron"	Burn & Co., Howrah.	
31651	Export	07/07/13	10t cd lt rr	"Baron"	Burn & Co., Howrah.	
31739	Export	13/10/13	4t sc rr	"Marquis"	Burn & Co., Howrah.	
31742	Export	27/10/13	10t sc rr	"Reading"	Burn & Co., Howrah.	
31801	Export	18/10/13	4t sc rr	"Marquis"	A.Henninger & Co. Darmstadt, Germany	
31818	Export	13/10/13	10t cd lt rr	"Baron"	Burn & Co., Howrah.	
31939	Export	06/02/14	10t sc rr	"King"	Gebr. Bobe GMBH, Dresden.	RH Steer
31956	Export	05/12/13	8t cd rr col	"Prince"	J.Chambers & Son Ltd., Auckland, NZ.	Fitted with differential gear and two tine Allen scarifier.
31992	Export	10/02/14	5/6t sc rr	"Marquis"	Burn & Co., Howrah.	
32076	Export	28/01/14	8t sc rr	"Ruler"	Java State Railways, via H.F.Roll.	
32111	AF4441	05/02/14	12t sc rr	"King"	R.Dingle & Sons, Stoke Climsland, Cornwall.	"Dunheved"
32119	Export	10/03/14	6t sc rr	"Monarch"	Karachi, via Burn.	Sold as 8 ton. /39 Spares to CE M&E dept., Jammu, Kashmir State.
32120	Export	07/04/14	9t sc rr	"Reading"	Burn & Co., Howrah.	
32137	Export	17/03/14	10t sc rr	"King"	A.Henninger & Co. Darmstadt, Germany	Bomford 2 tine scarifier
32138	Export	09/04/14	8t sc rr	"Ruler"	A.Henninger & Co. Darmstadt, Germany	Without roller rims.
32139	Export	23/05/14	8t sc rr	"Ruler"	A.Henninger & Co. Darmstadt, Germany	Without roller rims.
32224	Export	03/04/14	8t cd rr	"Prince"	Municipalidad de Cordoba, via Mattos.	
32255	Export	25/05/14	10t sc rr	"Reading"	Burn & Co., Howrah.	
32277	Export	19/05/14	6t sc rr	"Monarch"	Burn & Co., Howrah.	Sold as 8 ton.
32312	Export	16/06/14	5t sc rr col	"Earl"	Borneo Co.Ltd.	Samarang Exhibition.
32315	Export	10/07/14	18t cd rr	"Czar"	Municipalidad de Cordoba, via Mattos.	
32389	Export	15/07/14	10t sc rr	"King"	W&J Scheid, Limburg a.d. Lahn, Germany.	
32393	Export	29/10/14	4t sc rr	"Marquis"	Burn & Co., Howrah.	
32403	Export	30/10/14	7/8t sc rr	"Steel Earl"	Burn & Co., Howrah.	
32455	Export	29/10/14	12½t cd rr	"Special"	Frederiksborg County Council, via V.Andersen.	
32491	Export	23/10/14	8t cd rr	"Prince"	Mt. Albert Borough Council, via Chambers.	
32509		21/04/15	10t cd rr	"Autocrat"	Cork County Council.	
32522	Export	10/12/14	6t sc rr	"Monarch"	Burn & Co., Howrah.	Sold as 8 ton
32523	Export	27/02/15	10t sc rr	"Reading"	Burn & Co., Howrah.	
32585	Export	05/11/14	10t cd lt rr	"Baron"	Assiout Municipality, via T.Cook.	
32610	Export	28/12/15	10t sc rr	"King"	Borneo Co. Ltd.	
32621	Export	28/01/15	6t sc rr col	"Monarch"	Crown Agents for the Colonies.	Special, forks as "Reading"
32623	Export	26/11/14	10t cd lt rr	"Baron"	Colombo Electric Tramways & Lighting Co.Ltd.	Allen 3 tine scarifier.
32648	Export	25/02/15	8t cd rr col	"Prince"	Christchurch City Council Tramway Board.	via Macalister. Sprung front axle.
32665	Export	07/05/15	10t cd lt rr	"Baron"	Cavalla Town Council, Greece, via Gregoriades	
32731	Export	14/07/15	8t sc rr	"Monarch"	Burn & Co., Howrah.	
32737	Export	14/08/15	9t sc rr	"Ruler"	Czarnikow Rionda Co. New York.	Special with differential gear.
32779	Export	24/11/15	8t sc rr	"Ruler"	Java State Railways, via H.F.Roll.	
32802	Export	02/12/13	5/6t sc rr	"Marquis"	Burn & Co., Howrah.	
32922	Export	30/10/16	8t cd rr	"Prince"	Te Awamutu Borough Council, NZ.via Chambers.	
33021	Export	01/10/17	8t cd rr col	"Prince"	Royal State Rlys of Siam, Bangkok.	
33088	Export	18/09/17	14½t cd rr	"President"	Commission Internationale de Revitaillement.	
33089	Export	27/11/18	14½t cd rr	"President"	Commission Internationale de Revitaillement.	Shipped from Goole
33090	Export	27/11/18	14½t cd rr	"President"	Commission Internationale de Revitaillement.	Shipped from Goole
33631	Export	16/09/19	10t sc rr	"King"	J.Birch, Bombay, India	
33635		03/11/19	12t sc rr	"Queen"	Aitken, Thomson & Sons, Glasgow.	
33672	Export	22/10/19	8t sc rr col	Special	N.V.Soerabayasche Machinenhandel Voorheen	To Dutch East Indies Regs.via Becker&Co. Hague
33714	Export	22/10/19	8t sc rr col	Special	N.V.Soerabayasche Machinenhandel Voorheen Becke	To Dutch East Indies Regs.via Becker&Co. Hague
33731	Export	10/03/20	10t sc rr	"Reading"	J.Birch, Bombay, India	
33732	Export	10/03/20	10t cd rr	"Autocrat"	J.Birch, Bombay, India	
33783	Export	24/04/20	8t cd rr	"Prince"	J.Chambers & Son Ltd., Auckland, NZ.	Allen 2 tine scarifier.
33808	Export	09/07/20	5t sc rr	"Marquis"	J.Birch, Bombay, India	
33809	Export	13/05/20	7½t sc rr	"Earl"	J.Birch, Bombay, India	
33815	Export	12/08/20	10t sc rr	"King"	J.Birch, Bombay, India	
33856	Export	15/07/20	9t sc rr	"Ruler"	Turner, Hoare & Co, Bombay, via J.Birch.	
33857	Export	24/06/20	10t cd lt rr	"Baron"	J.Birch, Bombay, India	
33858	Export	15/07/20	12¾t cd rr	"Autocrat"	J.Birch, Bombay, India	
33859	Export	24/06/20	12t cd rr	"President"	J.Birch, Bombay, India	
33862	Export	26/07/20	9t sc rr	"Ruler"	J.Birch, Bombay, India	
33863	Export	26/07/20	9t sc rr	"Ruler"	J.Birch, Bombay, India	
34081	BJ6965	17/02/22	10t cd lt rr	"Baron"	RG&S, Leiston, Suffolk.	Supplied to E.H.Hunter, Yokohama.
34082	Export	11/03/22	10t cd lt rr	"Baron"	Kobe, Japan, via Hunter	
34083	Export	01/06/22	10t cd lt rr	"Baron"	Kobe, Japan, via Hunter.	
34084	BJ7045	03/02/22	10t cd lt rr	"Baron"	E.Suffolk C.C, Ipswich.	No.3
34085	BJ7044	08/02/22	10t cd lt rr	"Baron"	E.Suffolk C.C, Ipswich.	No.4, Scarifier fitted.
34165	PU1042	16/09/22	10t cd lt rr	"Baron"	Z.Fairclough, Clacton on Sea, Essex.	"Clacton Queen"
34166	Export	16/08/22	10t cd lt rr	"Baron"	T.Cook, Port Said.	
34173	Export	22/09/22	10t cd lt rr	"Baron"	Karachi, via Menzies.	Allen scarifier fitted.
34174	Export	22/09/22	10t cd lt rr	"Baron"	Karachi, via Menzies.	
34177		26/07/24	10t cd rr	"Autocrat"	Roscommon C.C.	No. 3
34185	Export	01/09/22	8t sc rr	"Ruler"	N.V.Soerabayasche Machinenhandel Voorheen	To Dutch East Indies Regs.via Becker&Co. Hague
34204	Export	10/11/22	10t cd lt rr	"Baron"	Min. Communications Main Roads&BridgesDept.	Egypt via T.Cook
34205	Export	10/10/22	10t cd lt rr	"Baron"	Bombay, via Macbeth Taylor	
34216	IE848	30/07/24	10t cd rr	"Autocrat"	Clare C.C.	/46 spares.
34217	Export	25/01/23	10t cd lt rr	"Baron"	La Casa Metzger S.A., Barcelona, Spain.	
34218		21/05/24	8t sc rr	"Ruler"	Roscommon C.C.	No. 2, Allen scarifier fitted.
34253	DL3031	05/03/23	10t sc rr	"Reading"	Borough of Newport, IOW.	/50 Spares.
34265	Export	14/02/23	10t cd lt rr	"Baron"	Julio Peña y Martin, Madrid.	Disused in Ibiza 1974, now at Long Shop Museum.
34266	Export	24/11/23	10t cd lt rr	"Baron"	Bombay, via R.Taylor.	
34267	BJ9844	27/10/24	10t cd lt rr	"Baron"	E.Suffolk C.C.	No.5, Scarifier fitted.
34585		29/10/24	10t cd rr	"Autocrat"	Clare C.C.	\46 spares
34594	IT644		10t cd rr	"Autocrat"	Leitrim C.C.	No.4
34608	IT6144	10/11/24	10t cd rr	"Autocrat"	Clare C.C.	\46 spares
34611		29/10/24	10t cd rr	"Autocrat"	Roscommon C.C.	No.4
34704	Export	03/10/27	10t cd rr	"Autocrat"	Gellatley Hankey for Hejaz, Arabia.	
34706	PU7494	31/03/25	10t cd lt rr	"Baron"	Z.Fairclough, Clacton on Sea, Essex.	"London Belle" 3 tine scarifier fitted.
34788	Export	15/09/25	8t cd rr col	"Prince"	Dundee Municipality, Natal, via Hubert Davies.	
34960	Export	25/02/27	4t sc rr col	"Marquis"	Jessop & Co., Rangoon.	
35131	Export	17/01/28	10t cd rr	"Autocrat"	Gellatley Hankey for Jeddah, Arabia.	

TRACTORS

Works No.	Regn. No.	Date	Type	Owner	Notes
25399	DX154	30/06/05	3 sc tr	R.J.Cole, Ipswich.	
25417	BJ372?	25/01/06	3 sc tr	H.Bennett, Stoke Ash, Suffolk.	Fore tank. Returned to RG&S
25480		01/08/05	3 sc tr	RG&S, Leiston, Suffolk.	Hull Show; Trial for Jones Haulage Syndicate; Hired to W.W.Howard Bros.
25565		20/01/06	3 sc tr	A.G.Cook, Haywards Heath, Sussex.	Belly tank, Returned to RG&S 12/10 part exchange for 4CD.
25646	Export	23/01/06	4 sc tr col conv	The Uganda Co.Ltd.	No. 1
25729	Export	15/03/06	3 sc tr	Grimaldi	Belly tank, Milan Exhibition.To RG&S 1/3/07 see 26580
25737	Export	23/02/06	4 sc tr	Bunger & Leyrer, Germany	To RG&S by 1/2/07 see 26515
25770	Export	08/08/06	4 sc tr	Burn & Co.	
25781		01/04/06	3 sc tr	RG&S, Leiston, Suffolk.	Copper top. Tavistock Show, see 27242
25794		09/05/06	3 sc tr	Parker Bros. (Mildenhall) Ltd., Suffolk.	
25795	Export	31/08/06	4 sc tr col	J.Macalister, Invercargill, NZ	
25888	DX225	24/07/06	4 sc tr	Bolton & Laughlin, Westerfield, Ipswich.	
25921	Export	27/06/06	3 sc tr	Alberto Casares	via Mattos. Wood block hind wheels. Awning over tender.
25957	L512	02/11/06	4 sc tr	L.Aston, Dinas, Glam.	RASE Derby.
26008	Export	01/11/06	4 sc tr col	Waygood Ltd., Sydney, N.S.W.	
26041	Export	16/11/06	4 sc tr col	Gilbert, Hamer & Co., Durban S.A.	
26063	BJ458	31/05/07	4 cd str	Jacob Studt, Southampton.	No dynamo platform. 3/09 to Clayton & Shuttleworth.
26432	Export	28/03/07	4 sc tr col	J.Macalister, Invercargill, NZ	
26433	Export	28/03/07	4 sc tr col	J.Macalister, Invercargill, NZ	
26479		25/07/07	4 cd tr	North Eastern Railway, Gateshead.	No awning.
26515	L514	25/04/07	4 sc tr	W&J.Jenkins Ltd., Brynsadler, Glam.	Was 25737
26518	OI599	06/09/07	4 cd tr	J.Harkness & Co., Belfast.	
26580		21/06/07	3 sc tr	E John, Smith & Co., Cowbridge,	Was 25729, awning over tender, returned to RG&S 1/08
26621	Export	20/08/07	4 sc tr	Burn & Co.	
26684	Export	02/10/07	4 sc tr col	J.Macalister, Invercargill, NZ	
26696	Export	25/01/08	4 sc tr col	J.Macalister, Invercargill, NZ	
26712	BJ533	11/11/07	4 cd tr	Lewis Brown, Headley, Newbury	
26719	Export	20/11/07	3 sc tr	J. Chambers & Son Ltd., Auckland NZ.	via Keep Bros. Birmingham. Awning over tender.
26722		14/02/08	4 cd tr	J.Wood & Son, Boughton Monchelsea, Kent.	
26969	Export	27/03/08	4 cd tr	Grimaldi, Milan, Italy.	
27050		18/02/08	4 cd tr	Lowcock Commercial Motor Co., Manchester.	In stock RG&S 31/12/08. Woodblock hind wheels.
27088	BJ561	31/05/08	4 cd tr	E.W.Rudd & Co., Poplar, London.	Woodblock hind wheels; demonstrator.
27108	BJ575	22/04/08	4 cd tr	Bricks Ltd., Newcastle upon Tyne.	
27112	Export	19/08/08	4 cd tr	Grimaldi, Milan.	Praienza Exhibition.
27163		01/07/08	4 cd tr	RG&S, Leiston, Suffolk.	Royal Show. No awning.
27164	BJ639	12/10/08	4 cd tr	Freston Bros., Ipswich.	Moore steam pump.
27170	Export	04/09/08	4 cd tr	Grimaldi.	
27183	Export	31/07/08	4 cd tr col	Ministerio Obras Publicas, B.A.	via Mattos
27185	BJ623	20/08/08	4 cd str	John Studt, Cardiff.	Returned to RG&S
27237	Export	21/09/08	3 sc tr	Grimaldi	Returned to RG&S 31/12/12
27242	Export	04/09/08	3 sc tr	Grimaldi	Was 25781. Awning over tender.
27243	Export	28/09/08	4 sc tr col	J.Macalister, Invercargill, NZ	
27274	Export	15/10/08	3 sc tr	Grimaldi	Awning over tender.
27342	BJ668	03/04/09	4 cd str	C.Hewson, Springburn, Glasgow.	"Pride of Scotland"
27351	Export	06/11/08	4 cd tr	J. Chambers & Son Ltd., Auckland NZ.	via Keep Bros. Birmingham.
27352	BJ563	11/12/08	4 cd tr	Coupe Bros., Sheffield.	
27356	BJ660	10/02/09	4 cd tr	W.Lennox, Belfast.	
27367	Export	19/02/09	4 cd tr col	Ministerio Obras Publicas, B.A.	via Mattos.
27368	Export	19/02/09	4 cd tr col	Ministerio Obras Publicas, B.A.	via Mattos.
27369	Export	19/02/09	4 cd tr col	Ministerio Obras Publicas, B.A.	via Mattos.
27370	Export	06/03/09	4 cd tr col	Ministerio Obras Publicas, B.A.	via Mattos.
27371	Export	19/02/09	4 cd tr col	Ministerio Obras Publicas, B.A.	via Mattos.
27372	Export	19/02/09	4 cd tr col	Ministerio Obras Publicas, B.A.	via Mattos.
27373	Export	06/03/09	4 cd tr col	Ministerio Obras Publicas, B.A.	via Mattos.
27374	Export	06/03/09	4 cd tr col	Ministerio Obras Publicas, B.A.	via Mattos.
27375	Export	06/03/09	4 cd tr col	Ministerio Obras Publicas, B.A.	via Mattos.
27376	Export	06/03/09	4 cd tr col	Ministerio Obras Publicas, B.A.	via Mattos.
27377	Export	03/04/09	4 cd tr col	Ministerio Obras Publicas, B.A.	via Mattos.
27378	Export	03/04/09	4 cd tr col	Ministerio Obras Publicas, B.A.	via Mattos.
27379	Export	03/04/09	4 cd tr col	Ministerio Obras Publicas, B.A.	via Mattos.
27380	Export	03/04/09	4 cd tr col	Ministerio Obras Publicas, B.A.	via Mattos.
27381	Export	03/04/09	4 cd tr col	Ministerio Obras Publicas, B.A.	via Mattos.
27620	Export	09/03/09	4 cd tr	Grimaldi to Bari.	RH Steer, equipped for direct ploughing.
27641	Export	05/04/09	4 cd tr	Pilter, Tunis.	
27655	BJ710	29/05/09	4 cd tr	J.Harkness & Co., Belfast.	
27673	BJ724	30/06/09	4 cd tr	F.O.Blaney, Lytchett Matravers, Dorset.	ex Royal Show, Poole.
27716	BJ728	06/07/09	4 cd tr	J.Harkness & Co., Belfast.	
27717		21/06/09	4 cd tr	County of Lanark.	Spares 3/23
27802	Export	17/08/09	4 cd tr col	J.Ballam, Camperdown, Natal.	Double draw bar. Lost at sea in SS Umhali.
27814	BJ727	29/07/09	4 cd tr	R.Corry Ltd., Belfast.	
27884	BJ769	28/10/09	4 cd str col	John Studt, Cardiff.	
27886	BJ750	07/09/09	4 cd tr	N.R.Etheridge, Blythburgh, Suffolk.	ex Yorkshire Show.
27912	Export	29/09/09	4 sc tr	J.Macalister, Invercargill, NZ	
27917	BJ782	13/04/11	4 cd tr	H.G.Brown, Leighton Buzzard, Beds.	"Nulli Secundus" Used in works for about 3 mths.
27931	Export	06/10/09	4 cd tr	Pilter, Algiers.	RH Steer, equipped for ploughing.
27937	BJ767	08/10/09	4 cd tr	R.Corry Ltd., Belfast.	
27941	Export	16/10/09	4 cd tr col	Ministerio Obras Publicas, B.A.	via Mattos.
27948	BJ783	14/12/09	4 cd tr	Colchester & Ball, Burwell, Cambs.	ex Smithfield Show.
27951	Export	20/12/09	4 cd tr col	J.Ballam, Camperdown, Natal.	Equipped for ploughing.
27960	BJ775	05/11/09	4 cd tr	J.Lucas, Holsworthy, Devon.	
28068	Export	01/03/10	4 cd tr sh col	Mattos	Buenos Aires Exhibition.
28076	Export	02/04/10	4 cd tr sh	E. Weber, Brussels.	Brussels Exhibition. 31/12/10 in stock
28079	BJ795	22/12/09	4 cd tr	A.F.Banks, Cardiff.	
28219	BJ792	29/12/09	4 cd tr	T.Collen J.P., Tanderagee, Co.Armagh.	
28220	BJ793	31/12/09	4 cd tr	A.Dawson & Co., Rushmere, Suffolk.	Waterheater
28258	BJ889	31/05/10	4 cd tr	S.Moore, Thornford, Sherborne, Dorset.	Devon Show.
28370	BJ799	19/02/10	4 cd tr	T.I.Mugleston, Stapleford Abbotts, Essex.	
28378	Export	21/03/10	4 cd tr	Vicomte de Sombreiul à L'isle Adam, France.	
28386	BJ885	25/06/10	4 cd tr	JP&W.Green, Drumore, Co.Down.	
28391	BJ904	09/08/10	4 cd tr	Saml.Rea, Antrim.	
28397	AH 048	28/07/10	4 cd tr	J.Walpole, Erpingham, Norwich.	ex Royal Show.
28404	Export	06/08/10	4 sc tr col	J.Macalister, Invercargill, NZ	
28409	BJ897	13/06/10	4 cd tr	J.Aird & Sons, Lambeth, London.	
28411	Export	08/10/10	4 cd tr	Alfred Devos, Brussels.	"La Courageuse"
28733	BJ1103	24/08/10	4 cd tr	R.Wynn & Sons, Newport, Mon.	Regn. BJ703 later in error.
28745	BJ987	21/12/10	4 cd tr	A.G.Cook, Haywards Heath, Sussex.	
28749	BJ978	12/11/10	4 cd tr	J.McIntosh, Forgue, Aberdeenshire.	"Mac O Forgue". No awning supplied.
28750	Export	26/11/10	4 sc tr col	J.Macalister, Invercargill, NZ	4 differential pinions.
28768	Export	25/01/11	4 cd tr sh col	Mattos	RH Steer and 'V' shaped strakes.
28774	Export	27/02/11	4 cd tr	H.Von Arx & Co., Zurich.	

Works No.	Regn. No.	Date	Type	Owner	Notes
28986	BJ1017	23/01/11	4 cd tr	Scutt & Sons, Blandford, Dorset.	Two injectors; no pump.
29005	Export	19/04/11	4 cd tr	Pilter, Marseille.	
29006	BJ1148	25/08/11	4 cd tr	H&J.Martin, Belfast.	Trailer brake gear; supplied with 5 ton trailer.
29013	Export	27/05/11	3 sc tr	Pilter, Dieppe.	Dynamo stand.
29017	Export	07/06/11	4 cd tr	Pilter, Marseille.	
29034	Export	02/08/11	4 cd tr	Pilter, Marseille.	
29767	Export	30/09/11	4 sc tr col	J.Macalister, Invercargill, NZ	
29771	BJ1162	07/09/11	4 cd tr	F.Matravers & Son, Coggeshall, Essex.	
29777	Export	21/09/11	4 cd tr col	J. Chambers & Son Ltd., Auckland NZ.	
29781	BJ1183	01/11/11	4 cd tr	H.Bennett, Stoke Ash, Suffolk.	"Fido"
29783	Export	30/09/11	4 cd tr sh col	J.Macalister, Invercargill, NZ	RH Steer, equipped for ploughing.
30007	BJ1257	15/03/12	4 cd tr	T.A.Everton Jr., Droitwich, Worcs.	
30050	BJ1189	24/11/11	4 cd tr	N.R.Etheridge, Blythburgh, Suffolk.	"George V"
30076	BJ1333	15/05/12	4 cd tr	J.Percy, Launceston, Cornwall.	
30085	BJ1317	30/04/12	4 cd tr	J.Brooks Taylor, Neath, Glam.	
30094	BJ1262	23/03/12	4 cd tr	D&C.Rutter, Crayford, Kent	Also at Gt.Wakering, Essex.
30140	Export	27/03/12	3 sc tr	Pilter, Marseille.	Open tender.
30143	Export	13/03/12	4 cd tr	Pilter, Marseille.	Open tender, no awning.
30159	BJ1206	23/12/11	4 cd tr sh	C.Purser, Cradley, Herefords.	
30858	BJ1368	08/07/12	4 cd tr sh	T.L.Walker, Shelsley Walsh, Worcs.	
30929	BJ1369	27/06/12	4 cd tr	Jones Bros., Treharris, Glam.	
30959	BJ1458	06/09/12	4 cd tr	J.Griffiths, Bargoed, Glam.	
30968	Export	31/12/12	4 cd tr	Pilter, Marseille.	Open tender, no awning.
30987	BJ1481	10/10/12	4 cd tr	J.Sanders & Sons, Hanley Castle, Worcs.	"Queen of Hanley Castle" Madresfield Show
30988	BJ1480	07/10/12	4 cd tr sh	G&J.Cutting, Pettaugh, Suffolk.	
31025	BJ1495	28/11/12	4 cd str	Thos. Cottrell, Leiston, Suffolk.	"East Lynne"
31108	BJ1502	18/11/12	4 cd tr	T.Evans, Troedyrhiw, Glam.	
31110	BJ1537	31/12/12	4 cd tr	J.James, Merthyr Tydfil, Glam.	via F.Munn, Cardiff.
31118	BJ1570	14/02/13	4 cd tr	Devon Trading Co.Ltd., Exeter.	3/22 part exchange for wagon
31126	Export	19/11/12	4 sc tr col	J.Macalister, Invercargill, NZ	
31173	BJ1520	12/12/12	4 cd tr	W.P.Williamson, Portadown, Co.Armagh.	
31193	BJ1659	01/05/13	4 cd tr	Warlow & Warlow, Newport, Mon.	
31209	BJ1514	31/12/12	4 cd tr sh	N.R.Etheridge, Blythburgh, Suffolk.	
31210	BJ1643	19/04/13	4 cd tr sh	R.Neill & Sons, Bangor, Co.Down.	
31248	BJ1559	22/02/13	4 cd tr sh	T.Moy, Colchester, Essex.	
31373	Export	22/02/13	4 cd tr	M.Sigart, Brussels.	
31435	BJ1654	29/04/13	4 cd tr	J.Snell, Bruisyard, Suffolk.	
31511	Export	05/06/13	4 cd str	Grand Manège de la Vallée d'Eure, France.	J.Broquet.
31632	BJ1726	10/07/13	4 cd tr sh	Rose & Sons, Kings Lynn, Norfolk.	
31633	BJ1696	28/06/13	4 cd tr	E.Clark, Hadleigh, Suffolk.	
31647	Export	01/08/13	4 cd tr sh col	James & Copeland.	High ground clearance, 201 gallon belly tank, RH steer.
31685	Export	20/06/13	4 cd tr	Jean Damoiseaux, Sittard, Holland.	Dutch regs., 6" extensions to rear wheels.
31704	BJ1758	29/07/13	4 cd tr sh	F.G.Bishop, Sible Hedingham, Essex.	
31705	Export	02/08/13	4 cd tr	M.Sigart, Brussels.	
31751	BJ1796	22/08/13	4 cd tr	Norfolk Motor Transport Co., Norwich.	Trailer brake gear.
31756	BJ1826	24/09/13	4 cd tr	S.P.Robinson, Stanton, Suffolk.	
31777	BJ1821	05/09/13	4 cd tr sh	Hough & Co.Ltd., Kings Heath, Birmingham.	
31780	BJ1814	09/09/13	4 cd tr sh	B.W.R.Hillen, Maldon, Essex.	
31800	BJ1837	07/10/13	4 cd tr sh	A&R.Brown, Colinsburgh, Fife.	
31804	BJ1846	18/11/13	4 cd tr	N.R.Etheridge, Blythburgh, Suffolk.	No. 4
31840	Export	12/11/13	4 cd str	Pilter, Paris.	
31895	BJ1851	14/11/13	4 cd tr	R.Dingle & Sons, Stoke Climsland, Cornwall.	
31901	BJ1858	08/11/13	4 cd tr sh	Express Motor Transport Co., Pershore, Worcs.	Express No. 5
31903	BJ1880	23/12/13	4 cd tr	A.King & Son, Wymondham, Norfolk.	
31926	BJ1870	28/11/13	4 cd tr	Cornish Tin Smelting Co., Redruth, Cornwall.	
31934	Export	31/01/14	4 cd tr sh	Bells United Asbestos Co.Ltd.	ex Smithfield Show. Exported to Kudaru, via Lagos, Nigeria.
31936	Export	10/12/13	4 cd tr	Pilter, Marseille.	Open tender, no awning.
31978	BJ1895	23/12/13	4 cd tr	R&J.Pierce, Belfast.	
31979	BJ1912	31/01/14	4 cd conv	G.Weller, Cranleigh, Surrey.	
31980	BJ1913	21/01/14	4 cd tr	F.J.Matthews, Abercarn, Mon.	
32029	BJ1894	15/03/14	4 cd tr sh	H.Lancaster & Co., Newent, Gloucs.	
32035	Export	24/12/13	4 cd tr sh col	R.Bartlett, Motupipe, Takoka, S.Island, NZ.	85 gallon belly tank; no awning.
32042	BJ1968	13/03/14	4 cd tr	Oxford Steam Plough Co., Cowley, Oxon.	
32048	BJ1969	25/02/14	4 cd tr	H.Cox, Yate, Gloucs.	
32049	Export	31/12/13	4 cd conv	Pilter, Paris.	
32084	BJ1943	09/03/14	4 cd tr	T.W.Ward Ltd., Silvertown, London.	Tangent Wheels.Tyres - hind wood & rubber, front steel.
32102	Export	04/03/14	4 cd tr	M.Sigart, Brussels.	
32110	BJ2062	20/04/14	4 cd tr	E.Holder, Kington Langley, Wilts.	
32122	BJ1911	24/01/14	4 cd tr	Jones Bros., Treharris, Glam.	
32144	Export	17/04/14	4 cd str	Pilter	RH steer. Customer supplied awning.
32190	BJ2063	18/04/14	4 cd tr	Devon Trading Co.Ltd., Exeter.	Returned to RG&S by 6/25
32200	BJ2064	29/04/14	4 cd tr	R&J.Pierce, Belfast.	
32295	BJ2067	09/05/14	4 cd tr sh	R.M'Lean, Aberfeldy, Perthshire.	"Mountaineer" No. 3. Water heater.
32335	BJ2061	08/06/14	4 cd tr	A.King & Son, North Mill, Wymondham, Norfolk.	No. 2
32357	BJ2070	25/06/14	4 cd tr	Norfolk Motor Transport Co., Norwich.	Trailer brake gear.
32358	BJ2076	17/06/14	4 cd tr sh	W.J.Last, Stonham Aspal, Suffolk.	
32360	BJ2201	04/07/14	4 cd tr sh	W.Aitken, Linlithgow, West Lothian.	Tubular waterheater.
32363	Export	05/05/14	4 cd str col	Pilter.	RH Steer. Customer supplied awning.
32379	BJ2203	07/07/14	4 cd tr sh	T.Evans, Ruabon, Denbighshire.	Tubular waterheater.
32415	BJ2066	17/06/14	4 cd tr	Curry & Dodd, Belfast.	
32473	BJ2078	15/10/14	4 cd tr sh	E.W.Rudd Ltd., Poplar, London.	Trailer brake gear.
32493	BJ2215	27/08/14	4 cd tr	J.Percy & Son, Launceston, Cornwall.	
32507	BJ2207	17/07/14	4 cd tr	T.A.Everton Jr. Droitwich, Worcs.	
32519	BJ2320	03/11/14	4 cd tr	Cork C.C.	
32524	Export	22/10/14	4 cd tr sh col	Crown Agents for the Colonies	Tubular waterheater; oil burner. Malaya.
32525	Export	22/10/14	4 cd tr sh col	Crown Agents for the Colonies	Tubular waterheater; oil burner. Malaya.
32533	BJ2219	12/09/14	4 cd tr	C.Blewett, Little Yeldham, Essex.	Double draw bar.
32556	Export	24/03/15	4 cd tr	Pilter	RH steer.
32559	BJ2327	11/11/14	4 cd tr	Cornish Tin Smelting Co., Redruth, Cornwall.	
32609	BJ2315	13/10/14	4 cd tr	Cork C.C.	
32615	BJ2321	29/10/14	4 cd tr sh	A.B.Goulbourne, Bidford on Avon, Warks.	Tubular waterheater.
32616	BJ2325	07/11/14	4 cd tr	E.W.Rudd Ltd., Poplar, London.	Trailer brake gear.
32622	BJ2329	21/11/14	4 cd tr	Southern Transport Co. Ltd., Brighton, Sussex.	No. 6. Tubular waterheater.
32631	BJ2395	19/12/14	4 cd tr	Coupe Bros., Sheffield.	
32632	BJ2391	17/12/14	4 cd tr	A.H.Williams, Monmouth.	
32633	BJ2398	16/01/15	4 cd tr	Wicklow C.C	No. 3
32634	BJ2484	02/03/15	4 cd tr	E.Packard & Co.Ltd., Ipswich.	
32637	BJ2393	29/12/14	4 cd tr sh	Lord Rayleigh, Terling, Essex.	ex Smithfield Show. Tubular waterheater.
32646	BJ2399	08/01/15	4 cd tr	T.Saul & Son Ltd., Gt.Yarmouth, Norfolk.	"Joyce"
32647	BJ2480	25/02/15	4 cd tr sh	Norfolk Motor Transport Co., Norwich.	No. 7
32653	BJ2408	08/02/15	4 cd tr	E&A.Shadrack, East Ham, London.	

Works No.	Regn. No.	Date	Type	Owner	Notes
32680	BJ2486	06/03/15	4 cd tr	Norman E.Box, Manchester.	"Bantam". Tangent hind wheels.
32681	BJ2493	25/03/15	4 cd tr	G.Beer, Alvington, Gloucs.	Tubular waterheater.
32682	BJ2488	01/04/15	4 cd tr	A.Lewis, Swaffham, Norfolk.	Lettered C.Rasberry , Gayton, Norfolk
32685	Export	13/05/15	4 cd col conv	J. Chambers & Son Ltd., Auckland NZ.	With Allen 3 tine scarifier.
32695	BJ2552	26/04/15	4 cd tr	Callan Co-op., Co.Kilkenny.	Tubular waterheater. Supplied with two 5/6 ton trailers.
32696	BJ2494	19/04/15	4 cd tr	Curry & Dodd, Belfast	
32697	BJ2560	07/06/15	4 cd tr	Pelabon Works, Richmond, Surrey.	Trailer brake gear. Supplied with four 5/6 ton trailers.
32703	BJ2562	20/05/15	4 cd tr	A.Thomas, Caerleon, Mon.	
32706	BJ2563	04/06/15	4 cd tr sh	ER&A.Debenham, Briantspuddle, Dorset.	
32716	BJ2556	25/05/15	4 cd tr	J.Davies, Redmarley, Gloucs.	"Queen of Hearts" . Tubular waterheater.
32724	BJ2672	30/07/15	4 cd tr	Cork C.C.	
32725	BJ2673	28/06/15	4 cd tr	Cork C.C.	
32730	BJ2569	26/06/15	4 cd tr	Fife Coal Co.Ltd., Leven, Fife.	Tubular waterheater. Supplied with three 5/6 ton trailers.
32738	BJ2674	08/07/15	4 cd tr	Austins Ltd., Ilford, Essex.	Supplied with three 6 ton trailers.
32739	BJ2681	15/07/15	4 cd tr	Bartlett Bayliss & Co.Ltd., Bideford, Devon.	For Coumbe Bros, Launceston
32740	BJ2558	14/05/15	4 cd tr	T.W.Ward Ltd., Silvertown, London.	Bauly Tangent Wheels.
32743	BJ2676	02/07/15	4 cd tr sh	J.Deans, Glenealy, Co.Wicklow.	Tubular waterheater.
32749	BJ2771	13/08/15	4 cd tr	J.George, Wetherden, Suffolk.	
32752	BJ2678	06/07/15	4 cd tr sh	T.A.Everton Jr., Droitwich, Worcs.	ex Royal Show. Tubular waterheater.
32753	BJ2688	05/18/15	4 cd tr	R.Dingle & Sons, Stoke Climsland, Cornwall.	
32760	BJ2770	10/08/15	4 cd tr	Hine Bros., Gillingham, Dorset.	"Sir John French"
32761	BJ2776	25/08/15	4 cd tr	C.Light & Co.Ltd., Christchurch, Hants.	
32762	BJ2774	20/08/15	4 cd conv	J.Cragg, Romsey, Hants.	
32765	BJ2777	01/09/15	4 cd tr	J.Thistleton Smith, Walsingham, Norfolk.	
32770	BJ2790	11/10/15	4 cd tr	Oxford Steam Plough Co., Cowley, Oxon.	Tubular waterheater.
32775	BJ2780	03/07/15	4 cd tr	H.Workman Ltd., Stroud, Gloucs.	Tubular waterheater.
32782	BJ2782	14/09/15	4 cd tr sh	Lord Rayleigh, Terling, Essex.	Lettered "Strutt & Parker, Southminster. Double draw bar, 6 ton trailer."
32783	BJ2853	14/10/15	4 cd tr	J.S.Baldock, Haverhill, Suffolk.	RH steer; double draw bar.
32784	BJ2860	27/10/15	4 cd tr	R.W.Christy, Boyton Hall, Chelmsford, Essex.	Double draw bar; 6 ton trailer.
32786	BJ2862	01/11/15	4 cd tr sh	J.Hickman & Sons, Wolverhampton.	
32789	BJ2865	12/11/15	4 cd tr	A.J.Hayward & Son, Colchester, Essex.	
32790	BJ2859	25/10/15	4 cd tr	G.V.Allen, York.	
32795	BJ2852	15/10/15	4 cd tr	W.Dunnett & Son, Woodbridge, Suffolk.	
32801	BJ2915	31/12/15	4 cd tr sh	A.Wilson & Sons, Dailly, Ayrshire	Tubular waterheater.
32809	BJ2866	27/11/15	4 cd tr	The Witham Cartage & Coal Co.Ltd., Essex.	
32813	BJ2913	18/12/15	4 cd tr	W.J.Overland & Son, Wisbech, Cambs.	Double draw bar.
32814	BJ2917	15/01/16	4 cd tr	F.Munn, Cardiff	
32815	BJ2916	31/12/15	4 cd tr	H.G.Cushion, Surlingham, Norfolk.	Double draw bar.
32819	BJ2920	11/01/16	4 cd tr	W.Vaughan, Newport, Salop.	Tubular waterheater. ex Smithfield Show.
32820	BJ2997	01/04/16	4 cd tr sh	E.G.French, Birdbrook, Essex.	Tubular waterheater.
32830	BJ2919	08/01/16	4 cd tr	Parker Bros. (Mildenhall) Ltd., Suffolk.	Tubular waterheater.
32834	BJ2929	28/02/16	4 cd tr	J.S.Baldock, Haverhill, Suffolk.	
32835	BJ3067	08/06/16	4 cd tr	B.T.Kent, Shelford Bottom, Cambs.	Double draw bar.
32836	BJ2991	03/03/16	4 cd tr	C&A.Knifton,Hertford Wharf, London.	Lettered "Henry Crane Cartage Contractor, Kingsland Basin"
32843	BJ2995	05/04/16	4 cd tr	Ashton Bros & Co.Ltd., Hyde, Cheshire.	Tubular waterheater.
32850	BJ3195	05/10/16	4 cd tr sh	Curry & Dodd, Belfast.	
32851	BJ3003	24/05/16	4 cd tr	Philipstown Co-op., Dunleer, Co. Louth.	IAWS Dublin.
32860	BJ3070	27/06/16	4 cd tr	R.Drake & Sons Ltd., Sutton, Cambs.	
32868	BJ3004	26/04/16	4 cd tr	W.S.Dewing, Norwich.	
32886	BJ3066	01/06/16	4 cd tr	E.W&W.F.Crane, Gt.Fransham, Norfolk.	
32913	BJ3074	11/05/16	4 cd tr	J.C.Edge & Co., Craven Arms, Salop.	
32923		20/10/16	4 cd col conv	Cornwall C.C.	No. 6
32924	BJ3200	30/10/16	4 cd tr	Thomas Briggs (Manchester) Ltd.	Tubular waterheater; trailer brake gear.
32933	BJ3197	07/10/16	4 cd tr	J.Sanders & Sons, Hanley Castle, Worcs.	"King of Hanley Castle"
32935	BJ3081	16/09/16	4 cd tr	A.Pipe, Sotterley, Suffolk.	
32944	BJ3206	17/11/16	4 cd tr	J.Abbott, Stowmarket, Suffolk.	
32945	BJ3268	23/12/16	4 cd conv	J.Wood & Sons, Boughton Monchelsea, Kent.	
32946	BJ3201	20/10/16	4 cd tr	E.C.Keith, Castleacre, Norfolk.	Later of Swanton Morley, Norfolk.
32950	BJ3272	05/01/17	4 cd tr	A.Lewis, Swaffham, Norfolk.	
32959	BJ3209	29/11/16	4 cd tr	Bowerman & Sons Ltd., Bridgwater, Somerset.	
32962	BJ3208	25/11/16	4 cd tr	W.Bartlett & Son, High Wycombe, Bucks.	
32969	BJ3280	16/02/17	4 cd tr	G.Davies & Son, Ludlow, Salop.	
32974	BJ3345	24/05/18	5 cd sh tr	S.J.Brown, Naas, Co.Kildare.	First DR tractor, completed June /17.
32978	BJ3276		4 cd tr sh	A.F.Wylie, Ardersier, Invernesshire.	"Cawdor Castle". Tubular waterheater.
32980	BJ3265	13/12/16	4 cd tr	Cullen, Allen & Co.Ltd., Belfast.	
32981	BJ3282	21/02/17	4 cd tr	Montgomeryshire Haulage Co., Newtown, Mont.	
32985	BJ3275	26/01/17	4 cd tr	Metropolitan Asylums Board, London.	Leavesden Asylum, Herts. Trailer brake gear.
32986	BJ3342	28/02/17	4 cd tr sh	Curry & Dodd, Belfast.	
33009	BJ3344	26/03/17	4 cd tr sh	P.McAinsh, Crieff, Perthshire.	"Shamrock" Sent to Dublin
33019	BJ3349	27/03/17	4 cd tr	Duchess of Hamilton, Easton, Norfolk.	Special draw bars.
33022	BJ3408	30/06/17	4 cd tr	W.McCready, Aughrim, Co. Wicklow.	Dept. of Agric. for Ireland.
33024	BJ3357	24/04/17	4 cd tr sh	F.B.Buxton, Bosley, Ches.	Tubular waterheater.
33025	BJ3358	24/04/17	4 cd tr	Sutton & Co., Guildford, Surrey	
33029	BJ3401	26/05/17	4 cd tr	War Office	Impressed by Impressing Officer's receipt dated 17/4/17
33032	BJ3393	08/05/17	4 cd tr sh	N.E.Box, Ardwick, Manchester.	Tubular waterheater. Rubber tyres.
33034	BJ3395	11/05/17	4 cd tr	War Office	Impressed.
33035	BJ3355	17/04/17	4 cd tr	M.Kirkpatrick & Sons, Athboy, Co.Meath.	
33037	BJ3398	23/05/17	4 cd tr	Earl of Ellesmere, Newmarket, Suffolk	
33043	BJ3353	10/04/17	4 cd tr	Central Wales Transport Co., Newtown, Mont.	
33052	BJ3409	23/06/17	4 cd tr	A.C.Hundy, Bricklehampton, Pershore, Worcs.	
33053	BJ3452	20/07/17	4 cd tr	W.H.Harley, Bridgnorth, Salop.	
33067	BJ3459	14/09/17	4 cd tr	Norfolk Motor Transport Co., Norwich.	No. 9. Tubular waterheater.
33074	BJ3451	12/07/17	4 cd tr	E.W.Howell, St.Weonards, Hereford.	Double draw bar.
33076	BJ3462	04/10/17	4 cd tr	Blackwater Valley Co-op., Co.Waterford.	Lettered "M.G.Cunningham, Cappoquin"
33078	BJ3410	13/07/17	4 cd tr	Metropolitan Asylums Board, London.	Joyce Green Hospital, Dartford.
33086	BJ3458	31/08/17	4 cd tr	S.Rea, Antrim.	
33087	BJ3460	15/09/17	4 cd tr	Outrath Co-op., Cahir, Co.Tipperary.	
33091	M-o-M	12/03/18	4 cd tr		
33107	BJ3454	13/08/17	4 cd tr	J.Taylor & Sons Ltd., Saffron Walden, Essex.	
33110	BJ3456	21/08/17	4 cd tr	J.P.Lawlor, Clondalkin, Co.Dublin.	
33111	BJ4145	18/10/18	4 cd tr	B.W.R.Hillen, Maldon, Essex.	
33116	BJ3461	24/09/17	4 cd tr	W.Wilson, Needham Mkt, Suffolk.	
33117	BJ3465	28/11/17	4 cd tr	Rathbritt Co-op. Implement Soc.Ltd., Co.Cork	Sunk in Irish Sea by U-Boat.
33123	BJ4165	31/12/18	4 cd tr	Montgomeryshire Haulage Co., Newtown, Mont.	
33127	BJ3464	14/11/17	4 cd tr	M.Kirkpatrick & Sons, Athboy, Co.Meath.	No. 2
33128	M-o-M	12/03/18	4 cd tr		
33131	BJ3468	13/02/18	4 cd tr	Tissara Co-op., Lisduff, Co.Roscommon.	
33132	BJ4148	31/10/18	4 cd tr sh	J.Barry & Sons Ltd., Fermoy, Co.Cork.	
33134	BJ3470	12/03/18	4 cd tr	M.Kirkpatrick & Sons, Athboy, Co.Meath.	No. 3
33141	M-o-M	27/02/18	4 cd tr		
33142	BJ4143	16/10/18	4 cd tr	T.B.Pink, Fareham, Hants.	"Edith Cavell"

Works No.	Regn. No.	Date	Type	Owner	Notes
33159	BJ3995	31/06/18	5 cd sh tr	F.W.D.Robinson, Beccles, Suffolk.	
33160	BJ4238	07/03/19	5 cd sh tr	Strutt & Parker, Southminster, Essex.	
33179	BJ4197	06/02/19	5 cd sh tr	A&H.Turner, Dunmow, Essex.	
33180	BJ4483	15/07/19	5 cd sh tr	J.Goddard & Son, Tunstall, Suffolk.	
33182	BJ4003	16/10/18	4 cd tr	L.Blakstad, Birmingham.	
33214	BJ3467	14/02/18	4 cd tr	P.O'Brien, Liscarroll, Co.Cork.	
33215	BJ4000	17/09/18	3 sc tr	Castlemahon Co-op., Co.Limerick.	Extended awning.
33219	BJ4164	28/12/18	4 cd tr	JW&A.Pearl, Otley, Suffolk.	"Vindictive" Reregd BJ6180 /21
33220	BJ4194	20/01/19	4 cd tr	E.G.French, Birdbrook, Essex.	"Bonny Jeanne"
33221	BJ4193	20/01/19	4 cd tr	Sutton & Co. (Guildford) Ltd., Surrey.	
33225	M-o-M	12/03/18	4 cd tr		
33236	BJ4161		4 cd tr	A.E.Allen, Tassagh, Co.Armagh.	
33242	M-o-M	12/03/18	4 cd tr		
33243	M-o-M	12/04/18	4 cd tr		
33244	BJ3850	13/03/18	4 cd tr	W.S.Dewing, Norwich.	
33246	BJ4883	08/03/20	5 cd sh tr	T.Francis, Fakenham, Norfolk.	
33247	Export	13/03/20	5 cd sh tr col	Sena Sugar Estates Ltd., Vila de Sena,	Portuguese East Africa.
33253	BJ4150	04/11/18	4 cd tr sh	McCue Dick & Co., Belfast.	
33258	BJ4196	28/02/19	4 cd tr	J.Hill, Manchester.	Tubular waterheater. Rubber tyres on hind wheels.
33259	BJ4159	31/10/18	4 cd tr	J.Porter & Son, Norwich.	
33263	BJ3859	07/05/18	4 cd tr	Coupe Bros., Sheffield	
33264	M-o-M	12/04/18	4 cd tr		
33265	BJ3852	18/03/18	4 cd tr	Carn Co-op., Curraghboy, Co.Roscommon.	
33266	BJ3853	14/03/18	4 cd tr	J.Russel, Mitchelstown, Co.Cork.	
33267	M-o-M	12/04/18	4 cd tr		
33269	M-o-M	09/05/18	4 cd tr		
33270	M-o-M	12/04/18	4 cd tr		
33271	M-o-M	26/04/18	4 cd tr		
33272	M-o-M	12/04/18	4 cd tr		
33273	M-o-M	26/04/18	4 cd tr		8" wide front wheels.
33274	M-o-M	26/04/18	4 cd tr		8" wide front wheels.
33277	M-o-M	26/04/18	4 cd tr		8" wide front wheels.
33278	M-o-M	22/04/18	4 cd tr		8" wide front wheels.
33282	BJ4158	09/11/18	4 cd tr	A.C.Hundy, Bricklehampton, Pershore, Worcs.	"Ambition"
33283	M-o-M	22/04/18	4 cd tr		8" wide front wheels.
33284	BJ4149	30/10/19	4 cd tr	Ipswich Steam Haulage Co.	
33285	BJ4240	26/03/19	4 cd tr	A.Sharman, Steeple Bumpstead, Essex.	
33288	M-o-M	09/05/18	4 cd tr		8" wide front wheels.
33289	M-o-M	09/05/18	4 cd tr		
33290	M-o-M	22/05/18	4 cd tr		
33292	M-o-M	22/04/18	4 cd tr		8" wide front wheels.
33293	M-o-M	22/04/18	4 cd tr		8" wide front wheels.
33294	M-o-M	22/05/18	4 cd tr		
33295	M-o-M	09/05/18	4 cd tr		
33296	M-o-M	22/04/18	4 cd tr		8" wide front wheels.
33300	M-o-M	13/07/18	4 cd tr		8" wide front wheels.
33301	M-o-M	17/06/18	4 cd tr		8" wide front wheels.
33303	BJ4151	13/11/18	4 cd tr	Saunders & Rowe, Chagford, Devon.	"Queen of the Moor"
33304	M-o-M	13/07/18	4 cd tr		8" wide front wheels.
33305	M-o-M	17/06/18	4 cd tr		8" wide front wheels.
33306	M-o-M	13/07/18	4 cd tr		8" wide front wheels.
33318	M-o-M	11/07/18	4 cd tr		8" wide front wheels.
33321	M-o-M	13/07/18	4 cd tr		8" wide front wheels.
33323	M-o-M	17/06/18	4 cd tr		8" wide front wheels.
33324	M-o-M	13/07/18	4 cd tr		8" wide front wheels.
33325	M-o-M	17/06/18	4 cd tr		8" wide front wheels.
33326	M-o-M	17/06/18	4 cd tr		8" wide front wheels.
33332	BJ3996	06/06/18	4 cd tr	Curry & Dodd, Belfast.	
33340	M-o-M	24/07/18	4 cd tr		8" wide front wheels.
33341	M-o-M	11/07/18	4 cd tr		8" wide front wheels.
33342	BJ4144	19/10/18	4 cd tr	Cornish Tin Smelting Co., Redruth, Cornwall.	
33343	BJ4142	16/10/18	4 cd tr	E.H.Jones, Newtown,	Tubular waterheater.
33344	M-o-M	13/07/18	4 cd tr		8" wide front wheels.
33345	M-o-M	11/07/18	4 cd tr		8" wide front wheels.
33346	M-o-M	24/07/18	4 cd tr		8" wide front wheels.
33347	M-o-M	23/08/18	4 cd tr		8" wide front wheels.
33348	M-o-M	23/08/18	4 cd tr		8" wide front wheels.
33349	M-o-M	23/08/18	4 cd tr		8" wide front wheels.
33350	M-o-M	24/07/18	4 cd tr		8" wide front wheels.
33351	BJ4160	02/12/18	4 cd tr	F.Flower Grimes, Purleigh, Essex.	
33353	BJ4236	25/02/19	4 cd tr	R.Dingle & Sons, Stoke Climsland, Cornwall.	"Prince"
33355	M-o-M	23/08/18	4 cd tr		8" wide front wheels.
33356	M-o-M	09/09/18	4 cd tr		8" wide front wheels.
33357	M-o-M	23/08/18	4 cd tr		8" wide front wheels.
33358	M-o-M	01/10/18	4 cd tr		8" wide front wheels.
33362	M-o-M	09/09/18	4 cd tr		8" wide front wheels.
33363	M-o-M	23/08/18	4 cd tr		8" wide front wheels.
33364	M-o-M	01/10/18	4 cd tr		8" wide front wheels.
33369	M-o-M	30/09/18	4 cd tr		8" wide front wheels.
33370	M-o-M	09/09/18	4 cd tr		8" wide front wheels.
33373	M-o-M	30/09/18	4 cd tr		8" wide front wheels.
33374	M-o-M	01/10/18	4 cd tr		8" wide front wheels.
33379	M-o-M	30/09/18	4 cd tr		8" wide front wheels.
33380	M-o-M	01/10/18	4 cd tr		8" wide front wheels.
33381	M-o-M	01/10/18	4 cd tr		8" wide front wheels.
33382	M-o-M	01/10/18	4 cd tr		8" wide front wheels.
33388	BJ3999	20/09/18	4 cd tr	Sir A.Herbert, Whitchurch, Hants.	
33395	M-o-M	25/10/18	4 cd tr		8" wide front wheels.
33396	M-o-M	25/10/18	4 cd tr		8" wide front wheels.
33397	M-o-M	18/11/18	4 cd tr		8" wide front wheels.
33398	M-o-M	18/11/18	4 cd tr		8" wide front wheels.
33399	M-o-M	20/12/18	4 cd tr		8" wide front wheels.
33400	M-o-M	25/10/18	4 cd tr		8" wide front wheels.
33401	M-o-M	18/11/18	4 cd tr		8" wide front wheels.
33402	M-o-M	16/12/18	4 cd tr		8" wide front wheels.
33403	M-o-M	31/01/19	4 cd tr		8" wide front wheels.
33404	M-o-M	16/12/18	4 cd tr		8" wide front wheels.
33405	M-o-M	18/11/18	4 cd tr		8" wide front wheels.
33406	M-o-M	20/12/18	4 cd tr		8" wide front wheels.
33407	M-o-M	20/12/18	4 cd tr		8" wide front wheels.
33408	M-o-M	20/12/18	4 cd tr		8" wide front wheels.

Works No.	Regn. No.	Date	Type	Owner	Notes
33410	M-o-M	20/12/18	4 cd tr		8" wide front wheels.
33411	BJ4297	25/04/19	4 cd tr sh	W&H.Middleton, Dennington, Suffolk.	
33412	BJ4307	22/05/19	4 cd tr sh	W.W.Kennedy & Co., Belfast.	
33414	BJ4195	28/01/19	4 cd tr	J.Harkness & Co., Belfast.	
33415	M-o-M	20/12/18	4 cd tr		8" wide front wheels.
33416	M-o-M	20/12/18	4 cd tr		8" wide front wheels.
33417	M-o-M	20/12/18	4 cd tr		8" wide front wheels.
33418	M-o-M	20/12/18	4 cd tr		8" wide front wheels.
33419	BJ4201	12/02/19	4 cd tr	J.Harkness & Co., Belfast.	
33422	BJ4162	07/12/18	4 cd conv	A.W.Maconochie, Cudham, Kent	
33423	BJ4198	31/03/19	4 cd tr	G&J.Glover, Tamnadoney, Co.Derry.	
33424	M-o-M	11/02/19	4 cd tr		8" wide front wheels.
33425	BJ4200	20/02/19	4 cd tr	W.G.Edney, Cosham, Hants.	"Denmead Champion"
33426	BJ4243	28/03/19	4 cd tr	Lancs. C.C., Preston.	No.1. Rubber tyres to hind wheels.
33427	BJ4199	27/02/19	4 cd tr	Gregor Bros., Aberdare, Glam.	
33430	BJ4235	20/02/19	4 cd tr	P.McAinsh, Crieff, Perthshire.	Fittings only for awning supplied.
33433	BJ4302	29/04/19	4 cd tr	E.C.Andrews, Stretton Court, Hereford.	
33434	BJ4237	20/03/19	4 cd conv	Earl of Leicester, Holkham, Norfolk.	Not superheated with piston valves.
33454	BJ4239		4 cd tr	F.R.Howe & Co., Ashcott, Somerset.	
33458	BJ4242	05/04/19	4 cd tr	Monmouth Steam Sawmills, Coal & Lime Co.Ltd.	
33459	BJ4305	05/05/19	4 cd tr	Lancs. C.C., Preston.	No.2. Rubber tyres to hind wheels.
33460	BJ4296	23/04/19	4 cd tr	Town & County Motor Garage Ltd., Aberdeen.	
33468	BJ4486	10/07/19	4 cd tr	Armstrong Addison & Co., Sunderland.	Tubular waterheater.
33485	BJ4294	10/04/19	4 cd tr	W.J.King, Bagboro', Somerset.	
33486	BJ4301	30/04/19	4 cd tr	Browning Bros., Stonehouse, Gloucs.	
33488	BJ4304	05/05/19	4 cd tr	J.B.Kind Ltd., Burton on Trent, Staffs.	
33492	BJ4299	16/04/19	4 cd tr	J.Wilks & Sons, Pontypool, Mon.	
33503	BJ4312	20/05/19	4 cd tr	N.E.Box Ltd., Ardwick, Manchester.	"Pilot" Rubber tyres to hind wheels.
33505	BJ4308	14/05/19	4 cd tr	Fletcher Bros., Birmingham & Cardiff.	"Rambler. Awning over smokebox. Solid flywheel."
33508	Export	13/03/20	5 cd sh tr col	Sena Sugar Estates Ltd., Vila de Sena,	Portuguese East Africa.
33510	BJ4306	14/05/19	4 cd tr	J.E.Corfield, Newtown, Mont..	Tubular waterheater.
33511	BJ4379	11/06/19	4 cd tr	J.E.Corfield, Newtown, Mont.	Tubular waterheater.
33513	BJ4309		4 cd tr	J.Taylor & Sons Ltd., Fulbourn, Cambs.	
33521	BJ4313	23/05/19	4 cd tr	Eastern Counties Haulage Co., Ipswich.	
33523	BJ4381	27/05/19	4 cd tr	Fletcher Bros., Newtown, Mont.	"Little Jim". Awning over smokebox. Solid flywheel.
33526	BJ4380	11/06/19	4 cd tr	T.A.Everton, Droitwich, Worcs.	"Cissy"
33538	BJ4522	14/10/19	4 cd tr	B.W.R.Hillen, Maldon, Essex.	
33545	BJ4384	23/06/19	4 cd tr	J.Harkness & Co., Belfast.	
33550	BJ4378	30/05/19	4 cd tr	Stowmarket Timber Co.Ltd., Suffolk.	
33566	BJ4395	07/07/19	4 cd tr	Browning Bros., Stonehouse, Gloucs.	
33585	BJ4376	07/07/19	4 cd tr	Gosbeck & Dist. Farmers Assocn., Suffolk.	"The Express"
33586	BJ4491	29/07/19	4 cd tr	G.B.Bull, Newport, Mon.	
33595	BJ4393	11/07/19	4 cd tr	W.Black & Son, Brechin, Angus.	
33596	BJ4385	30/06/19	4 cd tr	R.Jones, Bishopswood, Herefords.	
33608	BJ4487	25/07/19	4 cd tr	Thomas Merthyr Colliery Co.Ltd., Glam.	Waunwyllt Colliery.
33612	BJ4484	16/07/19	4 cd tr	H.Lancaster & Co., Newent, Glos.	
33614	BJ4390	23/06/19	4 cd tr	Barnard & Middleton, Badingham Brick Works, Suffo	
33615	BJ4511	14/10/19	4 cd conv	Marquis of Graham, Arran Estate, Brodick, Arran.	
33632	TB3740	12/09/19	4 cd tr	Ulverstone R.D.C., Lancs.	Rubber tyres to hind wheels.
33633	BJ4509	30/08/19	4 cd tr	S.Rea, Antrim.	
33636	BJ4514	20/09/19	4 cd tr	Ecclesiastical Commissioners, London.	Tubular waterheater.
33645	BJ4598	15/10/19	4 cd tr	W.I.Davies, Bucknell, Salop. (Ludlow Timber Co.)	
33646	BJ4505	13/09/19	4 cd tr	Garstang R.D.C., Lancs.	Rubber tyres to hind wheels.
33647	BJ4508	03/09/19	4 cd tr	H.Goddard & Son, Hasketon, Suffolk.	
33648	BJ4521	17/10/19	4 cd tr	F Munn, Cardiff.	Intended for J Davies, Redmarley, Gloucs.
33649	BJ4515	22/09/19	4 cd tr	Tidenham & Dist. Farmers Co-op., Gloucs.	
33656	BJ4519	26/09/19	4 cd tr	E.G.French, Birdbrook, Essex.	
33665	TB4890	31/10/19	4 cd tr	Fylde R.D.C., Lancs.	Rubber tyres to hind wheels.
33668	BJ4610	19/11/19	4 cd tr	J.Talbot, Lisduff, Co.Laois.	
33670	BJ4599	20/10/19	4 cd tr	Eddison Steam Rolling Co., Dorchester.	No. 278
33671	BJ4510	13/09/19	4 cd tr	Abergavenny R.D.C., Mon.	
33677	BJ4603	05/11/19	4 cd tr	J.Brien, Hazelhatch, Co.Dublin.	
33689	BJ4601	30/10/19	4 cd tr	J.Williamson, Worfield, Salop.	
33698	BJ4787	24/11/19	4 cd tr	Lancaster R.D.C.	
33705	BJ4788	03/12/19	4 cd tr	Stone Court Brick & Tile Works, Pembury, Kent.	
33706	BJ4875	29/01/20	4 cd tr	City of Cardiff P.W.Dept.	
33712	BJ4881	26/03/20	4 cd tr	J.White, Foxhill, Swindon, Wilts.	
33713	BJ4877	11/02/20	4 cd tr sh	N.E.Box, Ardwick, Manchester.	"Trojan". Rubber tyres to hind wheels. Tubular waterheater.
33715	BJ4789	12/12/19	4 cd tr	Curry & Dodd, Belfast.	
33716	BJ4872	12/01/20	4 cd tr	P.Last, South Green, Eye, Suffolk.	
33722		20/02/20	4 cd tr	IAWS (Dublin Stock).	
33738	BJ4791	08/12/19	4 cd tr	Eddison Steam Rolling Co., Dorchester.	No. 282
33739	BJ4803	06/01/20	4 cd tr	Eddison Steam Rolling Co., Dorchester.	No. 285
33741	BJ4800	08/01/20	4 cd tr	R.McKenzie, Ballygowan, Co.Down.	
33745	BJ4616	08/01/20	4 cd tr	A.J.King & Co.Ltd., Bollington, Macclesfield, Ches.	Rubber tyres to hind wheels.
33749	BJ5013	24/03/20	4 cd tr	Crichton Bros., Burrelton, Perths.	"Strathmore Champion"
33782	BJ4889	10/03/20	4 cd tr	W.Flowers & Sons, Pangbourne, Berks.	
33784	BJ5019	08/04/20	4 cd tr	Eddison Steam Rolling Co., Dorchester.	No. 286
33791	BJ5523		4 cd tr	B.W.R.Hillen, Maldon, Essex.	
33793	BJ5017	01/05/20	4 cd tr	W.Kayley, Ardwick, Manchester.	Rubber tyres to hind wheels.
33795	BJ5021	30/04/20	4 cd tr	W.J.Clayton Ltd., Saltley, Birmingham.	Rubber tyres to hind wheels.
33799	BJ4888	15/03/20	4 cd tr	Down C.C., Downpatrick.	No. 1. Six tipping and six non-tipping trailers with this order.
33800	BJ5148	13/05/20	4 cd tr	Down C.C., Downpatrick.	No. 2
33801	BJ5154	26/06/20	4 cd tr	Down C.C., Downpatrick.	No. 3
33802	BJ5018	01/04/20	4 cd tr	Down C.C., Downpatrick.	No. 4
33812	BJ5151	18/05/20	4 cd tr	Eddison Steam Rolling Co., Dorchester.	No. 287
33817	BJ5158	02/06/20	4 cd tr	J.L.Larkworthy, Worcester.	Worcester Show.
33818	BJ5323	14/08/20	4 cd conv	W.R.Odam, Thorney, S of Peterborough.	
33825	BJ5512	27/08/20	4 cd tr	G.Randle & Son Ltd., Bridgwater, Somerset.	
33826	BJ5331	21/07/20	4 cd tr	Stowmarket Timber Co.Ltd., Suffolk.	
33828	BJ5166	28/07/20	4 cd tr	J.Mellor & Sons, Guide Bridge, Manchester.	Rubber tyres, tubular waterheater.
33829	BJ5157	17/06/20	4 cd tr	C.Maidwell & Co.Ltd., Stowmarket, Suffolk.	
33838	BJ5342	23/08/20	4 cd tr	Blything R.D.C., Suffolk.	
33847	BJ5029	08/05/20	4 cd tr	R.H.Bunner, Montgomery.	
33881	BJ5332	30/07/20	4 cd tr	Holloway Bros (London) Ltd.	Rubber tyres
33882	BJ5335	16/07/20	4 cd tr	W.S.Dewing, Norwich.	
33899	BJ5333	14/08/20	4 cd tr	Curry & Dodd, Belfast.	
33902	BJ5340	20/08/20	4 cd str	J.Humphreys, Pelsall, Staffs.	"Medina"
33911	BJ5341	13/08/20	4 cd tr	K.Miles, Charminster, Dorset.	Trailer brake gear. Returned to RG&S 1926.
33941	BJ5513	27/08/20	4 cd tr	Metropolitan Asylums Board, London.	Darenth Training Colony, Dartford.
33949	ES3089	29/07/20	4 cd tr	Perthshire C.C.	

Works No.	Regn. No.	Date	Type	Owner	Notes
33950	ES3090	23/08/20	4 cd tr	Perthshire C.C.	
33952	YA232	01/11/20	4 cd conv	T.Bond, Kingston, Taunton, Somerset.	Flange on smokebox for A&P roller parts.
33953	BJ5522	11/09/20	4 cd tr	Long Ashton R.D.C., Somerset.	
33954	BJ5588	01/10/20	4 cd tr	H.Lancaster, Newent, Gloucs.	
33956	BJ5585	25/09/20	4 cd tr	G.Chamber, Sissinghurst, Kent.	
33980	BJ5595	26/10/20	4 cd tr	Bowerman & Sons Ltd., Bridgwater, Somerset.	
33981	BJ5598	06/11/20	4 cd tr	W.H.Gaze & Sons Ltd., Kingston on Thames, Surrey	
33982	RT125	09/07/23	4 cd tr	W.Last, Occold, Suffolk.	
33986	BJ5601	08/11/20	4 cd tr	E&H.Sadler, Stourbridge, Worcs.	
33987	BJ5597	29/10/20	4 cd tr	T&WJ.Hooper, Liskeard, Cornwall.	
33988		28/02/21	4 cd tr	Major Dermot McCalmont, Thomastown, Co.Kilkenn	
33989	BJ5603	26/11/20	4 cd tr	Ballymena U.D.C., Co.Antrim.	
33990	DN4581	29/07/21	4 cd tr	North Eastern Railway, Darlington, Co.Durham.	No. 108
33991	FK1300	12/08/21	4 cd tr	Martley R.D.C., Worcs.	
34005	NT589	05/09/21	4 cd tr	C.Norgrove, Bridgnorth, Salop.	
34006	BJ7831	30/12/22	4 cd tr	W.G.Baker, Halesworth, Suffolk.	"Excelsior"
34007	BJ8173	01/03/21	4 cd tr	AGE Dublin stock	2/10/23 to Herbert Hawes, Stradbroke, Suffolk.
34008	EC3766	23/08/21	4 cd tr	East Westmorland R.D.C., Kirkby Stephen.	
34019	KS1464	03/03/21	4 cd tr	Roxburgh C.C., Hawick District.	
34020	AF6080	01/11/22	4 cd tr	J.H.Parnell, Launceston, Cornwall.	"Pride of Cornwall"
34021	BJ7852		4 cd tr	RG&S, Leiston, Suffolk.	Yard tractor. 16/8/23 to Playle Bros. Heybridge, Essex.
34063	Export	31/03/21	4 cd col conv.	Carolus Brown, London for S.America.	
34134	CF4231	20/07/22	4 cd tr	H.N.Hensby, Newmarket, Suffolk	ex Royal Show.
34330	SE1761	30/11/23	4 cd conv	Banff C.C.	No. 3
34351		18/07/23	4 cd tr	Yandle & Sons, Martock, Somerset.	
34359	BJ8687	28/08/22	4 cd tr	H&B.Poole, Butley Abbey, Suffolk.	
34360	ER2659	05/07/24	4 cd tr	Watts & Son Ltd., Cambridge.	Second-hand ex works use.
34410	AN6214	12/02/24	4 cd tr	T.W.Ward Ltd., Silvertown, London.	Rubber tyres.
34461	PM4795	04/03/24	4 cd tr	W&F.Baines, Blackham, Sussex.	
34484	KI1116	02/07/24	4 cd conv	Waterford C.C.	
34503	Export	13/06/24	4 cd tr sh col	Balfour Lyon & Co., London for Chile.	Supplied with 6 Arato trailers.
34516	SE1979	30/07/24	4 cd tr	W.M.Robertson & Co., Banff.	
34517		16/07/24	4 cd tr	Roscommon C.C.	No. 3
34539	CF5488	30/06/24	4 cd tr	W.Bailey, Hengrave, Suffolk.	
34541		24/07/24	4 cd conv	Kilkenny C.C.	
34561	BJ9818	25/09/24	4 cd tr	S.R.Sherwood, Playford, Ipswich.	
34582		13/10/24	4 cd tr	P.G.C.Foster, Balsham, Cambs.	
34584		31/07/24	4 cd tr	Tipperary C.C., Nenagh.	Steam cock for taking steam to drill.
34592		/24	4 cd tr	RG&S, Leiston, Suffolk.	Works use.
34636	PU6267	11/02/25	4 cd tr	Witham Coal & Cartage Co.Ltd., Essex.	
34641	CF5913	/24	4 cd tr	RG&S, Leiston, Suffolk.	Smithfield Show. 6/4/25 to Clare RDC, Suffolk.
34684	PU6591	11/03/25	4 cd tr	BB&M.Barrell, Wivenhoe, Essex.	
34702	NR6159	31/03/25	4 cd tr	Coltman Bros, Rugby, Warks.	
34703		18/11/27	4 cd tr	Roscommon C.C.	¾" cock on boiler to drive jack hammers.
34713	FP1566?	28/04/25	4 cd tr	R&AJ.Rose, N.Luffenham, Rutland.	
34714	TW1636	16/02/26	4 cd tr	J.Brace & Sons, High Ongar, Essex.	
34729	DO5257	31/07/25	4 cd tr	S.Worth, Terrington St.Clement, Norfolk.	
34730	PW5878	11/08/25	4 cd tr	R.H.Holmes, Litcham, Norfolk.	
34776	DO5195	21/08/25	4 cd tr	A.Hovenden Worth, Fleet, Lincs.	
34787	AN8301	09/07/27	4 cd tr	T.W.Ward Ltd., Silvertown, London.	Rubber tyres; ex Norfolk Show.
34789	YB7841	15/11/26	4 cd tr	T.Parker & Son, Kingdon, Somerset.	
34855	Export	06/02/26	4 cd tr col	Ukanma Plantations Ltd., Kenya.	Direct ploughing parts.
35129	VG906	26/07/28	4 cd tr	A&W.Cushion, Norwich.	Rubber strakes.
35173	VG3251		4 cd tr	W.S.Dewing, Norwich.	
35225	OU3239	02/09/29	4 cd tr	W.Stanford, Preston Yeovil,	Rubber tyre segments on wheels.

Fig.149. *The last word! The final 4CD tractor No.34225, delivered on September 2, 1929, to W. Stanford of Preston, Yeovil, Somerset, now owned by Andrew Hawkswell of Sheriff Hutton, and seen here in 1992 at Hovingham in a timeless shot, makes a fitting conclusion to this final instalment of the history of Garrett self-moving engines. The firm and much of the Works have gone, but enough of their engines survive to ensure that they will not be forgotten.* [A. Hawkswell]